Topics in modern mathematics 1

Topics in modern mathematics 1

T. D. H. Baber

Ph.D., M.Sc., B.Sc., Dip. Ed., F.I.M.A.
Principal of Farnborough Technical College

Pitman Publishing

First published 1972

SIR ISAAC PITMAN AND SONS LTD.
Pitman House, Parker Street, Kingsway, London, WC2B 5PB
P.O. Box 6038, Portal Street, Nairobi, Kenya

SIR ISAAC PITMAN (AUST.) PTY. LTD.
Pitman House, 158 Bouverie Street, Carlton, Victoria 3053, Australia

PITMAN PUBLISHING COMPANY S.A. LTD.
P.O. Box 11231, Johannesburg, S. Africa

PITMAN PUBLISHING CORPORATION
6 East 43rd Street, New York, N.Y. 10017, U.S.A.

SIR ISAAC PITMAN (CANADA) LTD.
495 Wellington St West, Toronto 135, Canada

THE COPP CLARK PUBLISHING COMPANY
517 Wellington St West, Toronto 135, Canada

A/510

Cased edition: ISBN 0 273 31679 6
Paperback edition: ISBN 0 273 31681 8

Made in Great Britain at the Pitman Press, Bath
G2-(T.373/1360: 75)

Preface

Over recent years, the approach to the teaching of mathematics has changed considerably in recognition of the need to provide a more effective programme of mathematical education in schools and colleges. There is general agreement that curricula and teaching methods require modernization. Reforms have been proposed by various organizations with the aims of promoting an early understanding of the basic structure of mathematics and of eliminating outmoded traditional material.

In various modern programmes, attention has been concentrated upon algebra, analysis and geometry treated in a more advanced and abstract way than formerly. The traditional approach, by which these subjects were compartmentalized, has been discarded and the present aim is to achieve a unified approach by which the relationships between these subjects are fully exploited and the underlying unity of mathematics is exhibited. It has been deemed desirable and practicable to introduce topics, formerly reserved for more advanced courses, at an earlier stage, e.g. the concepts and language of set theory are now widely recognized as providing an excellent medium for promoting the understanding and appreciation of a wide range of mathematical topics.

This volume presents a number of modern topics suitable not only for students who are proceeding to the study of mathematics, science or engineering but also as a programme of general education on the reasonable assumption that modern mathematics can make an important contribution in the field of liberal education. Whilst the presentation of these topics is non-traditional, students, whose earlier mathematical education has proceeded on traditional lines, should experience no handicap.

Volume I is based upon a series of lectures given by the author a few years ago to all undergraduates in the University of Malawi, irrespective of their specializations, to introduce them to modern mathematical thinking and teaching as a form of liberal education. The author makes no claim for comprehensiveness though the basic topics of mathematics and certain interesting and important applications of the subject have been included in this volume.

The two volumes, comprising this textbook, have not been written to cover any specific syllabuses. It is, however, claimed that they contain logically developed expositions of the more important topics of modern mathematics which have gained prominence in a wide range of examination syllabuses.

Volume 1 deals with the structure of the real number system and their representation on the number line. The study of inequalities, ordered pairs, relations and functions is developed in terms of set theory which is itself considered in an early chapter. A chapter on non-decimal arithmetic is followed by an account of digital computers and the elements of programming. This volume concludes with chapters on linear programming and an introduction to matrices and vectors. This volume is suitable for Vth and VIth formers in secondary schools, particularly for those who are studying modern mathematics syllabuses and also for students in Technical Colleges who are pursuing O.N.D./C. courses in Engineering and Science into which modern topics are being increasingly introduced.

Volume 2 contains subject matter of a more advanced standard including vector differentiation and integration, probability theory, Boolean algebras, and group theory. Since these topics are included in H.N.D./C. and degree courses in Mathematics, Science and Engineering, this volume is suitable for students pursuing such courses in Technical Colleges, Polytechnics and Universities.

In conclusion, I wish to thank Dr. Lee Peng Yee of the University of Malawi and Mr. R. W. Boxer of Farnborough Technical College for their helpful comments and suggestions.

T. D. H. Baber

Contents

1 The system of real numbers

1.1 Introduction

The four basic operations and their symbols,

addition (+)   subtraction (−)   multiplication (. or ×)   division (÷)

as applied to ordinary numbers such as 2, −5, 3/4 and $\sqrt{3}$, are very familiar to all students. These numbers are members of the system of *real numbers*.

Most students take for granted the truth of such statements as the following:

$$(-3) \times (-2) = 6 \qquad \frac{10}{-2} = -5 \qquad \frac{2}{15} = \frac{10}{75} \qquad \frac{3}{4} \div \frac{4}{5} = \frac{15}{16}$$

These results are by no means obvious and require careful justification.

In this Chapter, we shall examine the nature of the underlying structure of the system of real numbers and justify the validity of such operations upon members of the system.

1.2 Basic Postulates

The use of letters, such as a, b, x and y, to achieve greater generality is well known. If we represent any two numbers by a and b, their sum is represented by $a + b$ and their product by $a \cdot b$, $a \times b$, or simply ab. Thus addition of any two numbers a and b leads to another number, their sum $a + b$, and their multiplication also leads to another number, their product *ab*. We refer to such operations, whereby two numbers are combined to form a third, as *binary operations*.

Members of the system of real numbers are numbers which, under the binary operations of algebra, obey a group of rules or postulates.

The basic postulates for the system of real numbers follow. Let x, y, z be any real numbers, then they obey the following laws:

POSTULATE 1:   *The Closure Property*
The sum $x + y$ and the product xy are themselves real numbers and are unique.

POSTULATE 2:   *The Commutative Laws of Addition and Multiplication*

$$x + y = y + x$$

$$xy = yx$$

POSTULATE 3:   *The Associative Laws of Addition and Multiplication*

$$(x + y) + z = x + (y + z)$$

$$(x . y) . z = x . (y . z)$$

POSTULATE 4:   *The Distributive Law*

$$(x + y) . z = x . z + y . z$$

POSTULATE 5:   *The Existence of Identity Elements*
0 and 1 are respectively identity elements relative to addition and multiplication, i.e.

$$0 + x = x + 0 = x$$

$$1 . x = x . 1 = x$$

POSTULATE 6:   *The Existence of an Additive Inverse*
$(-x)$ is the additive inverse of x, i.e.

$$x + (-x) = (-x) + x = 0$$

and $(-x)$ is a unique real number.

POSTULATE 7:   *The Existence of a Multiplicative Inverse*
x^{-1} is the multiplicative inverse of x provided $x \neq 0$, i.e.

$$x . x^{-1} = x^{-1} . x = 1$$

and x^{-1} is a unique real number.

Postulate 1 expresses the fact that, under the binary operations of addition and multiplication, any two real numbers lead to a unique sum and a unique product which themselves belong to the real number system. This is the property of *closure* and we say that the real number system is *closed* under addition and multiplication.

We know by experience that whether we add 4 to 3 or 3 to 4, we obtain the sum 7. Similarly we know that $3 . 4 = 4 . 3 = 12$. Postulate 2, which gives the commutative laws, expresses these facts in general terms.

Again it is a matter of experience that the order in which we add or multiply three numbers is immaterial. This is expressed formally in the associative laws of Postulate 3, which imply that the sum and the product of three real numbers x, y and z may be written, without ambiguity, as $x + y + z$ and xyz respectively, omitting the brackets.

These laws are readily extended to more than three numbers, e.g.

$$3 + 4 + 6 + 7 = (3 + 4) + 6 + 7 = (3 + 4) + (6 + 7)$$
$$= 7 + 13 = 20$$

$$3 . 4 . 6 . 7 = (3 . 4) . 6 . 7 = (3 . 4) . (6 . 7)$$
$$= 12 . 42 = 504$$

The numbers 3, 4, 6 and 7 are called *terms* of the sum 20 and are called *factors* of the product 504.

The distributive law of Postulate 4 may be written, in reverse, as

$$x . z + y . z = (x + y) . z = z . (x + y)$$

showing that the factors of the sum on the left are z and $(x + y)$.

The distributive law may be extended thus:

$$a(x + y + z) = a[(x + y) + z] = a(x + y) + az = ax + ay + az$$
$$(a + b)(x + y) = a(x + y) + b(x + y) = ax + ay + bx + by$$

a and b being any real numbers.

In Postulate 5, we define two real numbers, represented by the symbols 0 and 1, which are called *additive* and *multiplicative identities* respectively, such that the addition of 0 and multiplication by 1 leave any real number unchanged.

Postulate 6 states that any real number x is associated, under addition, with a corresponding unique real number $(-x)$ called its additive inverse or *negative;* and similarly Postulate 7 states that under multiplication, any real number x is associated with a corresponding unique real number x^{-1} called its multiplicative inverse or *reciprocal* provided that x is different from 0.

1.3   *Subtraction*

By Postulate 6, $(-x)$, the negative of any real number x, is itself a real number and

$$x + (-x) = 0$$

By the same postulate, since $(-x)$ is a real number, it will have a negative which must be denoted by $[-(-x)]$ and

$$[-(-x)] + (-x) = 0$$

It therefore follows that

$$[-(-x)] = x$$

Thus

$$[-(-2)] = 2 \quad \text{and} \quad [-\{-(-2)\}] = -2$$

and we derive the simple rule that, if a real number x is preceded by an odd number of negative signs, its value is $-x$ but if preceded by an even number of negative signs, its value is x.

DEFINITION 1.1
We define the *subtraction* of y from x as the *addition* of $-y$ to x, i.e.

$$x - y = x + (-y)$$

It will be noted that the symbol $-$ is used to denote both the operation of subtraction and the negative of a number.

The number $z = x - y$ is called the *difference* of x and y. It follows that

$$
\begin{aligned}
z + y &= (x - y) + y \\
      &= [x + (-y)] + y \\
      &= x + [(-y) + y] \quad &\text{(Postulate 3)} \\
      &= x + 0 \quad &\text{(Postulate 6)} \\
      &= x \quad &\text{(Postulate 5)}
\end{aligned}
$$

Thus $z = x - y$ implies that $z + y = x$ and vice versa.
The result $-(x + y) = -x - y$ follows from Postulates 2, 3, 5 and 6.

EXAMPLE 1.1
Assuming only the Definition 1.1 for subtraction and the conclusions which
have been derived from it above:
(A) Simplify $-4 + 9 - 6$.

$$\text{Let } x = -4 + 9 - 6$$
$$= 9 - 4 - 6$$
$$= 9 - (4 + 6)$$
$$= 9 - 10$$
$$(-x) = 10 - 9$$
$$9 + (-x) = 10$$

Since 1 must be added to 9 to give 10, $(-x) = 1$, i.e. $x = -1$.
(B) Subtract $[-3 - (-7)]$ from -8.

$$\text{Let } x = -8 - [-3 - (-7)]$$
$$= -8 - (-3 + 7)$$
$$= -8 - (-3) - 7$$
$$= -(8 + 7) + 3$$
$$= -15 + 3$$
$$(-x) = -3 + 15$$
$$3 + (-x) = 15$$

Therefore $(-x) = 12$ and $x = -12$.

1.4 Multiplication

In this section, we shall consider how the negatives of real numbers and the
number 0 behave under the operation of multiplication in accordance with
the basic postulates.

(i) To PROVE that $x \cdot 0 = 0$.*

$$
\begin{aligned}
0 &= xy - xy & \text{(Postulate 6)} \\
&= x(y + 0) - xy & \text{(Postulate 5)} \\
&= xy + x \cdot 0 - xy & \text{(Postulate 4)} \\
&= xy - xy + x \cdot 0 & \text{(Postulate 2)} \\
&= 0 + x \cdot 0 & \text{(Postulate 6)} \\
&= x \cdot 0 & \text{(Postulate 5)}
\end{aligned}
$$

It follows that if $xy = 0$ either $x = 0$ or $y = 0$.

(ii) To PROVE that $x(-y) = -xy$

$$0 = x \cdot 0 = x[y + (-y)] = xy + x(-y)$$

But $0 = xy - xy$ and since every real number has a unique additive inverse, it follows that

$$x(-y) = -xy$$

Similarly, it may be proved that

$$(-x)y = -xy \quad \text{and} \quad (-x)(-y) = xy$$

(iii) To PROVE that $(xy)^{-1} = x^{-1}y^{-1}$ provided $x \neq 0$ and $y \neq 0$.

$$
\begin{aligned}
(xy)(x^{-1}y^{-1}) &= xyx^{-1}y^{-1} & \text{(Postulate 3)} \\
&= xx^{-1}yy^{-1} & \text{(Postulate 2)} \\
&= 1 \cdot 1 & \text{(Postulate 7)} \\
&= 1 & \text{(Postulate 5)}
\end{aligned}
$$

But $(xy)(xy)^{-1} = 1$ and since any real number has a unique multiplicative inverse, it follows that

$$(xy)^{-1} = x^{-1}y^{-1}$$

EXAMPLE 1.2
(A) Multiply $(x - a)$ by $(y + b)$ showing the essential steps of the process.

$$
\begin{aligned}
(x - a)(y + b) &= x(y + b) + (-a)(y + b) \\
&= xy + xb + (-a)y + (-a)b \\
&= xy + bx - ay - ab
\end{aligned}
$$

* It is usually considered to be obvious that $x \cdot 0 = 0$ but this statement must not be accepted without proof.

(B) Factorize $2ax - 6bx - 3ay + 9by$.

$$2ax - 6bx - 3ay + 9by = 2x(a - 3b) - 3y(a - 3b)$$
$$= (a - 3b)(2x - 3y)$$

1.5   Division

Subtraction has been defined in terms of the operation of addition. In a similar way, division may be defined in terms of the operation of multiplication.

DEFINITION 1.2
We define the *division* of x by y ($y \neq 0$) as the *multiplication* of x by y^{-1}
That is

$$x \div y = \frac{x}{y} = xy^{-1}$$

On putting $x = 1$, we have $y^{-1} = 1/y$, i.e. y^{-1} is the reciprocal of y.
$z = x/y$ is called the *quotient* of x by y. The symbol x/y is called a *fraction*. The number x is called the *numerator* and y is called the *denominator*.

$$zy = \left(\frac{x}{y}\right)y = xy^{-1} y = x$$

Thus $z = \frac{x}{y}$ implies $zy = x$.

It will be seen that division by 0 has been excluded.
If $y = 0$ then $yz = 0$ for all values of z. Hence no unique value of the quotient $z = x/y$ is possible, i.e. z would not be defined when $y = 0$. Consequently, division by 0 is not defined.
We now prove that the reciprocal of the reciprocal of a non-zero number is the number itself.

(i) To PROVE that $(x^{-1})^{-1} = x$ ($x \neq 0$)
The reciprocal of the number x^{-1} is $(x^{-1})^{-1}$. Therefore

$$x^{-1}(x^{-1})^{-1} = 1$$

But $x^{-1}x = 1$ and since the reciprocal of a non-zero real number is unique, it follows that

$$(x^{-1})^{-1} = x$$

In Section 1.3, it was proved that $-(-x) = x$, so that it has now been proved that the inverse of the inverse of a number, under both addition and multiplication, is the number itself.

(ii) To PROVE that $-x^{-1} = (-x)^{-1}$ $(x \neq 0)$

$$(-x)(-x^{-1}) = xx^{-1} = 1 \qquad \text{since} \quad (-a)(-b) = ab$$

It follows that $-x^{-1}$ is the reciprocal of $-x$, i.e. $-x^{-1} = (-x)^{-1}$.

(iii) To PROVE that $\dfrac{x}{-y} = -\dfrac{x}{y}$

$$\frac{x}{y} + \frac{x}{-y} = x[y^{-1} + (-y)^{-1}]$$

$$= x[y^{-1} - y^{-1}] \qquad \text{by (ii) above}$$

$$= x \cdot 0$$

$$= 0$$

Since x/y, is a real number, it will have a negative $-x/y$, so that

$$\frac{x}{y} + \left(-\frac{x}{y}\right) = 0$$

Since the negative of any real number is unique, it follows that

$$\frac{x}{-y} = -\frac{x}{y}$$

Similarly it may be proved that

$$\frac{-x}{y} = -\frac{x}{y}$$

EXAMPLE 1.3

(A) Find the quotient $\dfrac{16}{-2}$.

Let $z = \dfrac{16}{-2}$. Then

$$(-2)z = 16$$

(-2) must be multiplied by (-8) to give 16. Therefore $z = -8$.

(B) Divide -3 by 12.

Let $z = \dfrac{-3}{12}$.

$$12z = -3$$
$$12zz^{-1} = -3z^{-1}$$
$$12 = -3z^{-1}$$

Also $12 = (-3)(-4)$.

By comparison of the factors of 12, it follows that

$$z^{-1} = -4$$
$$z = \dfrac{1}{-4} = -\dfrac{1}{4}.$$

(C) Divide $px - qy + qx - py$ by $x - y$.

$$px - qy + qx - py = px + qx - qy - py$$
$$= x(p + q) - y(p + q)$$
$$= (x - y)(p + q)$$

Let z be the quotient. Then

$$z = \dfrac{px - qy + qx - py}{(x - y)} \qquad \text{provided } x - y \neq 0$$
$$= \dfrac{(x - y)(p + q)}{(x - y)}$$
$$z(x - y) = (p + q)(x - y)$$
$$z = p + q$$

EXERCISES 1.1

1. State the best order of performing the following operations mentally
   (i) $47 + 268 + 13$ (ii) $154 + 72 + 18$ (iii) $-135 + 69 + 336$
   (iv) $25 . 78 . 4$ (v) $6 . 137 . 5$

2. Evaluate
   (i) $7(3 + 5)$ (ii) $7 . 3 + 5$ (iii) $(3 + 4)6 + 5$ (iv) $3 + 4(6 + 5)$
   (v) $(3 + 4)(6 + 5)$ (vi) $3(8 - 4) + 6$ (vii) $3 . 8 - (4 + 6)$

3. Simplify the following, justifying each step by quoting the postulates or results used:
   (i) $-3 + 8 - 6$ (ii) $7 - [(-4) + 6]$ (iii) $-4(-5 + 9)$
   (iv) $-[4 - (-8)(5 - 9)]$ (v) $(a - 2b)(2x + y)$ (vi) $(-3a + 2b)(2x - 3y)$
   (vii) $(-2a)(-b)(3c) - a(-2b)(-c)$ (viii) $(x - 7)(y + 3)$

4. Factorize

    (i) $(x - a)y - (-x - b)y$ (ii) $(y - a)z - (x - 3a)(-z)$

    (iii) $ax + ay + bx + by$ *(iv)* $ax - 2ay + 3bx - 6by$

5. Divide the following, justifying each step in the process:

    (i) $-10 \div 5$ (ii) $-4 \div 8$ (iii) $2^{-1} \div 6^{-1}$ (iv) $(14 - 2) \div 9$

    (v) $1 \div \dfrac{x^{-1}}{y^{-1}}$ (vi) $(ab - ac) \div (b - c)$

State the restrictions which must be imposed in (v) and (vi).

6. Prove the following results carefully stating the postulates used as in Section 1.4

    (i) $-(x + y) = -x - y$ (ii) $(-x)y = -xy$ (iii) $\dfrac{-x}{y} = -\dfrac{x}{y}$.

1.6 The Genesis of the Real Numbers

(a) THE INTEGERS

When the need to count his possessions became apparent, man was stimulated to invent a symbolism for enumerating them. The symbols $1, 2, 3, 4, 5, \ldots$, the "counting numbers", comprise such a collection of symbols and represent an early stage in the development of the real number system. The numbers $1, 2, 3, 4, 5, \ldots$ are known as the *natural numbers*.

Since the sum and the product of any pair of natural numbers are also natural numbers, it is clear that the system of natural numbers is closed under the operations of addition and multiplication. It is easily verified that the natural numbers obey the Distributive Law and the Commutative and Associative Laws for addition and multiplication and also have the multiplicative identity 1 in accordance with the postulates of Section 1.2 for the real numbers.

When the binary operations of addition and multiplication *only* are applied *within* the system of natural numbers, the number 0 does not arise. Furthermore, the multiplicative inverse of any natural number, except 1, would not be another natural number and consequently would have no meaning. Similarly, the negative of any natural number would not be a natural number and would likewise have no meaning. It follows that Postulate 5 (except for the multiplicative identity), and Postulates 6 and 7 of the real number system would not be relevant to simple binary operations within the system of natural numbers.

The need for negative numbers arose in the process of subtraction, for example when any natural number was subtracted from a smaller natural number. If we are required to solve the equation $x + 7 = 3$, we seek a number which when added to 7 gives 3. Of course, there is no answer in terms of the natural numbers.

To meet this deficiency, a new kind of number, the negative whole number or *negative integer*, was invented. The negative integers are written -1, -2, $-3, \ldots$ and the numbers 1, 2, 3, $\ldots$ may also be written $+1$, $+2$ $+3, \ldots$ and are called *positive integers*.

Let us now identify -1, -2, $-3, \ldots$ with the additive inverses (-1), (-2), $(-3), \ldots$ respectively and, in general, let the negative integer $-N$ be identified with the inverse $(-N)$ of the natural number or positive integer N.

The *integers* $\ldots -3$, -2, -1, 0, 1, 2, 3, $\ldots$ comprise the positive and negative integers together with 0 which has been included to provide the additive identity of Postulate 5. The system of integers satisfies Postulate 6 and is closed under subtraction as well as under addition and multiplication. The integers obey all the postulates of the real number system except Postulate 7 concerning the multiplicative inverse. In general, the multiplicative inverse of an integer is not another integer and as such has no meaning within the system of integers.

(b) THE RATIONAL NUMBERS

Common fractions are to be found in the earliest mathematical records and arose from practical problems involving division. A new kind of number was required to express the value of the quotient when any integer was divided by a larger integer, e.g. no integer can be found to satisfy the equation $4x = 3$. The system of integers was therefore extended by the invention of *rational numbers*.

DEFINITION 1.3
A *rational number* is defined to be a number of the form p/q where p and q are integers and $q \neq 0$.

Since any rational number p/q is an integer if $q = 1$, the integers are included in the system of rational numbers.

If x and y are integers, the fraction x/y is called a simple or *common fraction*. Thus the rational numbers may be identified with the common fractions, for example

$$\frac{1}{8}, \frac{3}{4}, -\frac{2}{3}, \frac{9}{7} \text{ are rational numbers}$$

Using Definition 1.2 for division, we shall prove a number of results concerning the manipulation of common fractions.

(i) To PROVE that if $p/q = r/s$, then $ps = qr$ ($q \neq 0$, $s \neq 0$), and conversely.

$$pq^{-1} = rs^{-1}$$

Therefore

$$pq^{-1}qs = rs^{-1}qs = rs^{-1}sq$$
$$ps = qr$$

(The proof of the converse will be left to the student.)

It follows that $p/q = pr/qr$ and consequently that factors common to the numerator and the denominator of any common fraction may be cancelled. When all the factors common to the numerator and the denominator have been cancelled, i.e. when the numerator and the denominator have ultimately become relatively prime, the fraction is said to be reduced to its *lowest terms*. For example

$$\frac{45}{105} = \frac{9}{21} = \frac{3}{7}$$

so that the fraction 45/105 becomes 3/7 when reduced to its lowest terms.

It is clear that any common fraction may be written in many ways, e.g.

$$\frac{2}{3} = \frac{4}{6} = \frac{6}{9} = \cdots$$

(ii) To PROVE that $\dfrac{p}{q} \cdot \dfrac{r}{s} = \dfrac{pr}{qs}$

$$\frac{p}{q} \cdot \frac{r}{s} = pq^{-1} \cdot rs^{-1} = prq^{-1}s^{-1} = pr(qs)^{-1} = \frac{pr}{qs}$$

$$\frac{3}{4} \cdot \frac{2}{9} = \frac{6}{36} = \frac{1}{6}$$

(iii) To PROVE that $\dfrac{p/q}{r/s} = \dfrac{ps}{qr}$.

$$\frac{p/q}{r/s} = \frac{pq^{-1}}{rs^{-1}} = \frac{pq^{-1}qs}{rs^{-1}qs} = \frac{psq^{-1}q}{qrs^{-1}s} = \frac{ps}{qr}$$

i.e. division by r/s is equivalent to multiplication by s/r. For example

$$\frac{2/5}{7/11} = \frac{2 \cdot 11}{5 \cdot 7} = \frac{22}{35}$$

(iv) To PROVE that $\dfrac{p}{q} + \dfrac{r}{q} = \dfrac{p+r}{q}$.

$$\frac{p}{q} + \frac{r}{q} = pq^{-1} + rq^{-1} = q^{-1}(p + r) = \frac{p+r}{q}$$

Thus, in order to add or subtract two fractions, we replace each fraction by an equivalent fraction in such a way that each replacement fraction has the same denominator and then apply the above result. For example,

$$\frac{5}{6} + \frac{2}{15} = \frac{75}{90} + \frac{12}{90} = \frac{87}{90} = \frac{29}{30} \qquad (90 = 6 \times 15)$$

The arithmetic will be simplified if the common denominator used is the Lowest Common Multiple (LCM) of the denominators of the fractions. In the above example, the LCM of 6 and 15 is 30, and we have

$$\frac{5}{6} + \frac{2}{15} = \frac{25}{30} + \frac{4}{30} = \frac{29}{30}$$

Similarly in subtraction,

$$\frac{5}{6} - \frac{2}{15} = \frac{5}{6} + \frac{-2}{15} = \frac{25}{30} + \frac{-4}{30} = \frac{25 - 4}{30} = \frac{21}{30} = \frac{7}{10}$$

Similarly, algebraic fractions may be added or subtracted by finding the LCM of their denominators and then replacing each fraction by an equivalent fraction having the LCM as its denominator. For example,

$$\frac{1 - x}{x(1 + x)} - \frac{x}{1 - x^2} \qquad\qquad \text{LCM} = x(1 - x^2)$$

$$= \frac{(1 - x)(1 - x)}{x(1 - x^2)} - \frac{x^2}{x(1 - x^2)} = \frac{1 - 2x + x^2 - x^2}{x(1 - x^2)} = \frac{1 - 2x}{x(1 - x^2)}$$

We now know that the rules for addition and multiplication of any two rational numbers p/q and r/s are as follows:

$$\frac{p}{q} + \frac{r}{s} = \frac{ps + qr}{qs} \quad \text{and} \quad \frac{p}{q} \cdot \frac{r}{s} = \frac{pr}{qs}$$

Since $p/q + (-p/q) = 0$, any rational number p/q has an additive inverse $(-p/q)$ which may be identified with

$$-\frac{p}{q} = \frac{-p}{q} = \frac{p}{-q}$$

$(p/q)^{-1}$, the multiplicative inverse of any rational number p/q $(p \neq 0)$, is q/p since

$$\frac{p}{q}\left(\frac{p}{q}\right)^{-1} = 1 \quad \text{and} \quad \frac{p}{q} \cdot \frac{q}{p} = 1$$

0 and 1 are identity elements with respect to addition and multiplication respectively, since

$$\frac{p}{q} + 0 = \frac{p}{q} \quad \text{and} \quad \frac{p}{q} \cdot 1 = \frac{p}{q}$$

It may now be readily verified that the system of rational numbers, which includes the integers, is closed under addition, subtraction, multiplication and division and obeys *all* the postulates of Section 1.2 for the real number system.

There remains one more class of the real numbers, the *irrational numbers*, which will now be considered.

(c) IRRATIONAL NUMBERS

The rational numbers are adequate for everyday use in measurement. The dimensions of a building plan would be stated in rational multiples of feet or metres; the current in an electrical circuit would be measured in rational multiples of amperes; and the power of engines in rational multiples of the horsepower or kilowatt and so on.

From the mathematical point of view, the rational number system has certain shortcomings. For example, the diagonal of a unit square and the circumference of a unit circle cannot be expressed in rational multiples of any unit of length.

By the theorem of Pythagoras, the length of the diagonal of a unit square is $\sqrt{2}$. It may be proved that $\sqrt{2}$ is not a rational number, i.e. that $\sqrt{2}$ cannot be represented by a number p/q where p and q are integers. We use an indirect method of proof. We assume that $\sqrt{2}$ is rational and show that this assumption leads to a contradiction.

Let $p/q = \sqrt{2}$ where p/q is reduced to its lowest terms. Then

$$\frac{p^2}{q^2} = 2 \quad \text{i.e.} \quad p^2 = 2q^2$$

Since $2q^2$ is divisible by 2, p^2 is divisible by 2. Thus, p is divisible by 2 so that p is an even number.

Write $p = 2n$ where n is an integer.

$$4n^2 = 2q^2$$

or

$$2n^2 = q^2$$

Since $2n^2$ is divisible by 2, q^2 is divisible by 2, and therefore q is divisible by 2.

We have now shown that both p and q are divisible by 2 which contradicts our original assumption that p/q is in its lowest terms. It follows that $\sqrt{2}$ cannot be a rational number.

Any number, such as $\sqrt{2}$, which cannot be expressed as the quotient of two integers is called an *irrational number*.

The complete system of real numbers comprises the rational numbers and irrational numbers together and it may easily be verified that the family of real numbers as a whole obey *all* the postulates of Section 1.2 and form a closed system under the binary operations of addition, subtraction, multiplication and division.

TABLE 1.1   The family tree of the real numbers

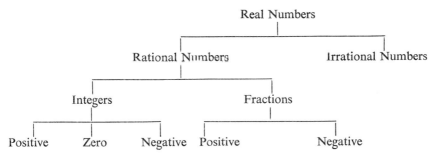

Table 1.1 illustrates the sub-classes of the system of real numbers. A most important property of the real number system is that its members may be placed in order. Ordering will be discussed in Chapter 5.

1.7   Rational Numbers as Decimals

Any rational number may be expressed in the decimal notation by the process of division.

It is found that the decimal equivalents of rational numbers are either

terminating decimals, e.g. $\dfrac{3}{4} = 0.75$

or recurring decimals, e.g. $\dfrac{1}{6} = 0.1666\cdots = 0.1\dot{6}$ and

$\dfrac{3}{7} = 0.428571\,428571\cdots = 0.\dot{4}2857\dot{1}$ (see Example 1.4B)

(A dot over a single digit implies that this digit is repeated indefinitely and two dots indicate the ends of a group of digits which is repeated indefinitely.)

Let us now prove that any rational number may be expressed as either a terminating decimal or a recurring decimal.

Let a/b be any rational number such that $-1 < a/b < 1$. (We may impose this restriction on the rational number, without loss of generality, since a rational number may always be expressed in the form of a mixed number.)

To express a/b as a decimal, we perform the division of a by b. If the process of division terminates, then a/b has been expressed as a terminating decimal. If the process does not terminate, then there will be a remainder less than b after each stage of division. There are, at most, $(b-1)$ possible remainders, namely $1, 2, 3, \ldots, (b-1)$. Therefore, after $(b-1)$ stages of division *at most*, the remainders must begin to repeat. Thus the digits in the quotient will repeat and a/b will be a recurring decimal.

EXAMPLE 1.4

(A) Express $-3/11$ as a decimal.

```
11)30(0.27
   22
   ──
   80
   77
   ──
   30
```
$$-\frac{3}{11} = -0.\dot{2}\dot{7}$$

(B) Express $3/7$ as a decimal.

```
7)30(0.428571
  28
  ──
  20
  14
  ──
  60
  56
  ──
  40
  35
  ──
  50
  49
  ──
  10
   7
  ──
  30
```
$$\frac{3}{7} = 0.\dot{4}2857\dot{1}$$

Now consider the reverse process.

Let us convert a recurring decimal into the equivalent rational number. Let the decimal contain a repeating group of r digits, $a_1 a_2 a_3 \ldots a_r$, each of the a's being one of the digits $1, 2, 3, \ldots, 9$.

Let x be the recurring decimal. Therefore

$$x = 0 \cdot a_1 a_2 a_3 \ldots a_r \, a_1 a_2 a_3 \ldots a_r \, a_1 \ldots$$
$$10^r x = a_1 a_2 a_3 \ldots a_r \cdot a_1 a_2 a_3 \ldots a_r \, a_1 a_2 a_3 \ldots a_r \, a_1 \ldots$$
$$x(10^r - 1) = a_1 a_2 a_3 \ldots a_r \cdot 0$$
$$x = \frac{a_1 a_2 a_3 \ldots a_r}{(10^r - 1)} \qquad \text{(a rational number)}$$

EXAMPLE 1.5

Express $0.\overline{153846}$ as a rational number.

Let $x = 0.153846 \, 153846 \, 1 \ldots$

Therefore

$$10^6 x = 153\,846 \cdot 153\,846 \, 153\,846 \, 1 \cdots$$
$$x(10^6 - 1) = 153\,846$$
$$x = \frac{153\,846}{999\,999} = \frac{17\,094}{111\,111} = \frac{5\,698}{37\,037} = \frac{814}{5\,291} = \frac{74}{481} = \frac{2}{13}$$

We conclude that every decimal which neither terminates nor recurs represents an *irrational number*.

1.8 Algebraic Structure and the Concept of Group

In this chapter, we have discussed the properties of the set* of real numbers, the elements of which combine under the operations of addition and multiplication according to the rules stated in Section 1.2.

The system of real numbers is an example of an *algebraic structure*.

DEFINITION 1.4

An *algebraic structure* is defined as a *set* of elements which are combined under specified *operations* according to stated rules.

One of the most interesting algebraic structures is the *group*. Let us consider the simplest kind of group, namely, a group for which only *one* operation is defined on elements of the set.

---

* The precise algebraic meaning of the term "set" will be discussed in Chapter 4.

DEFINITION 1.5
A set, upon which only one operation is defined, will be a group provided
  (i) the set is closed under the operation
  (ii) the operation is associative
  (iii) there is an identity element
  (iv) every element has an inverse.
It may easily be verified that the set of integers, Z, is a group under the operation of addition.
  (i) Closure: when two integers are added we obtain another integer:

$$5 + (-3) = 2 \quad \text{(Postulate 1)}$$

  (ii) The associative law for addition:

$$(5 + 3) - 4 = 4 = 5 + (3 - 4) \quad \text{(Postulate 3)}$$

  (iii) Identity element for addition:

$$4 + 0 = 4$$
$$-5 + 0 = -5 \quad \text{(Postulate 5)}$$

  (iv) Inverse under addition:

$$6 + (-6) = 0 \quad \text{(Postulate 6)}$$

The inverse of 6 is $(-6) = -6$, a negative integer.

$$(-3) + 3 = 0$$

The inverse of (-3) is 3, a positive integer.
  All the necessary conditions for a group are therefore satisfied by the integers under addition.
  The set of integers do not form a group under multiplication because condition (iv) is not satisfied for this operation, since no element, except 1, has a multiplicative inverse which belongs to the set. For example, $4 \times 1/4 = 1$ but 1/4 the multiplicative inverse of 4 does not belong to the set of integers.
  It may readily be verified, however, that the set of rational numbers and the set of real numbers are both groups under multiplication. In particular, we note that all multiplicative inverses are members of the respective sets.
  The theory of groups, initiated by Cauchy, has been one of the most fundamental and fruitful developments in algebra. The importance of this study lies in the fact that the *underlying structure* of one and the same group is often the essence of several apparently unrelated theories. A single investigation will then result in the solution of several problems.
  Further consideration will be given to the important topics of sets and groups later in this book.

EXERCISES 1. 2

1. Perform the operations indicated

(a) $\dfrac{7}{5} - \dfrac{2}{3}$ (b) $\dfrac{\dfrac{5}{7} - \dfrac{3}{5}}{\dfrac{4}{5} + \dfrac{3}{7}}$

(c) $\dfrac{x}{y} - \dfrac{y}{x}$ (d) $\dfrac{1}{x} + \dfrac{1}{y} + \dfrac{1}{z}$ (e) $\left(\dfrac{1}{x} - \dfrac{x}{2}\right)\left(\dfrac{x}{3} - \dfrac{1}{x}\right)$

(f) $\dfrac{1-x}{x} + \dfrac{x}{1+x}$ (g) $\dfrac{x+2y}{x^2+xy} + \dfrac{x-y}{xy+y^2}$

2. Show that $\sqrt{3}$ and $\sqrt{5}$ are irrational numbers by using the indirect method proof used in Section 1.6.

3. Express the following rational numbers as decimals indicating which terminate and which recur.
   (a) $\frac{5}{8}$ (b) $2\frac{3}{8}$ (c) $\frac{2}{3}$ (d) $22/7$ (e) $2/11$ (f) $\sqrt{(16/25)}$
   [(a) 0.675   (b) 2.375   (c) $0.\dot{6}$ (d) $2.\dot{1}4285\dot{7}$ (e) $0.\dot{1}\dot{8}$ (f) 0.8]

4. Express the following decimals as rational numbers, where possible
   (a) 0.3175   (b) 1.059   (c) $0.\dot{3}$ (d) $0.\dot{3}\dot{6}$ (e) $0.47\dot{2}$
   (f) $0.\dot{5}7142\dot{8}$ (g) 0.11010010001 . . .   (h) 0.525522555222 . . .

   $\left[\text{(a)} \dfrac{127}{400} \quad \text{(b)} \dfrac{1059}{1000} \quad \text{(c)} \dfrac{1}{3} \quad \text{(d)} \dfrac{4}{11} \quad \text{(e)} \dfrac{472}{999} \quad \text{(f)} \dfrac{4}{7} \quad \begin{array}{l}\text{(g) irrational} \\ \text{(h) irrational}\end{array}\right]$

5. If x is a positive integer, which is not a multiple of 10, show, using an indirect method of proof, that $\log_{10} x$ is an irrational number.
   [*Hint:* Show that the assumption $\log_{10} x = p/q$ leads to a contradiction of the hypothesis that x is not a multiple of 10.]

A/510

2 Non-decimal arithmetic

2.1 Arithmetic Bases

The numbers which are used in ordinary arithmetic, are expressed as multiples of powers of 10, e.g.

37 means $3 \times 10 + 7$

285 means $2 \times 10^2 + 8 \times 10 + 5$

This method of representing numbers has been in use for hundreds of years and is called the *common*, *denary*, or, more usually, *decimal* notation. The number 10 is called the *base* of the decimal number system. In this system, the symbols used are the digits $0, 1, 2, \ldots 9$.

Any number, other than 10, may also be used as the base of a number system. Thus, in octal arithmetic (base 8), only the digits $0, 1, 2, \ldots 7$ are used. In this system, we have, for example:

	Octal	*Decimal Equivalent*
75 means	$7 \times 8 + 5$	61
436 means	$4 \times 8^2 + 3 \times 8 + 6$	286

More generally, in arithmetic base r, the number

$$a_n r^n + a_{n-1} r^{n-1} + \cdots + a_2 r^2 + a_1 r + a_0$$

is represented by

$$a_n a_{n-1} \cdots a_2 a_1 a_0$$

where the digits $a_n, \ldots a_0$ are integers which are all less than r, of which one or more, excluding a_n, may be zero.

In arithmetic to any base greater than 10, additional symbols will be needed to represent digits greater than 9. Since arithmetic for bases greater than 12 will not be considered here, it is necessary to use only three additional symbols which will be t, e, and T^* representing 10, 11 and 12 respectively.

Binary arithmetic, base 2, using the two digits 0 and 1 only, has become very important since it is widely employed in digital computers which are well suited to operation in the binary system. An electronic computer is essentially a complex network of electrical switches each of which will be in one of two electrical states only—current is either flowing through the switch or it is not—corresponding to the binary digits 1 or 0 respectively. In essence, a computer is a device for performing addition and subtraction in the binary system. Multiplication is carried out by repeated addition and division by repeated subtraction.

EXAMPLE 2.1
(A) Add 110011 to 1101010 in the binary system. Subtract the first number from the second. Check the answers in the decimal system.

```
  1101010          1101010
+  110011        −  110011
─────────        ─────────
 10011101           110111
```

Binary   Decimal

$1101010 = 2^6 + 2^5 + 2^3 + 2 = 106$

$110011 = 2^5 + 2^4 + 2 + 1 = 51$

$10011101 = 2^7 + 2^4 + 2^3 + 2^2 + 1 = 157 = 106 + 51$

$110111 = 2^5 + 2^4 + 2^2 + 2 + 1 = 55 = 106 − 51$

In the above, note that

(a) in the second column from the right in the addition, $1 + 1 = 10$ so that 0 is recorded in the second column and 1 is added in the third column;

(b) in the first column on the right in the subtraction, $0 − 1 = 10 − 1 − 10 = 1 − 10$ so that 1 is recorded in the first column and 1 is subtracted in the second column.

---

* Although in arithmetic (base 12) the number twelve would be expressed as 10, it is nevertheless convenient, as will be seen later, to use the symbol T for twelve for certain purposes.

(B) Add 145263 to 634251 in the scale of eight. Also subtract the first number from the second.

$$
\begin{array}{r}
634251 \\
+145263 \\
\hline
1001534
\end{array}
\qquad
\begin{array}{r}
634251 \\
-145263 \\
\hline
466766
\end{array}
$$

(Note that in the subtraction 3 cannot be subtracted from 1. We therefore add 8 to 1 and take 3 from nine giving 6. We then take 7 from 8 + 5 giving 6 and so on.)

(C) Find the product of 8072 and 5316 in the scale of nine.

$$
\begin{array}{r}
5316 \\
8072 \\
\hline
11633 \\
41426 \\
46743 \\
\hline
47270003
\end{array}
$$

(On multiplying by 2, 2 × 6 = twelve = 1 × 9 + 3 so that 3 is recorded and 1 is carried and so on.)

(D) Find the square of *eee* in the scale of twelve.

$$
\begin{array}{r}
eee \\
eee \\
\hline
tee1 \\
tee1 \\
tee1 \\
\hline
ee\,t001
\end{array}
$$

(Note: *e* × *e* = one hundred and twenty one = 10 × 12 + 1)

(E) Divide 102432 by 36 in the scale of seven.

$$
\begin{array}{r}
36)102432(1625 \\
36 \\
\hline
334 \\
321 \\
\hline
133 \\
105 \\
\hline
252 \\
252 \\
\hline
\end{array}
$$

(F) In Binary arithmetic, multiplication and division are very simple.
(*a*) Multiply 110101 by 1011
(*b*) Divide 1111101 by 101

(*a*)

$$110101 = 53 \text{ (decimal)}$$
$$1011\ \ = 11 \text{ (decimal)}$$
$$\overline{110101}$$
$$110101$$
$$\underline{110101}$$
$$\overline{1001000111} = 583 \text{ (decimal)}$$

(*b*) 101)1111101(11001
$$\underline{101}$$
$$\overline{101}$$
$$\underline{101}$$
$$\cdots 101$$
$$\underline{101}$$

Check: $101 = 5$ (decimal)
$1111101 = 125$ (decimal)
$11001 = 25$ (decimal)

2.2 *Change of Base*

Let N be any integral number which is to be expressed in terms of a different base r. Let the n digits which represent N in the new scale be

$$a_n a_{n-1} \cdots a_1 a_0 \quad \text{in that order}$$

so that

$$N = a_n r^n + a_{n-1} r^{n-1} + \cdots + a_1 r + a_0$$

$a_0, a_1, a_2, \ldots, a_n$ are to be found.
On dividing N by r, the quotient is

$$a_n r^{n-1} + a_{n-1} r^{n-2} + \cdots + a_1$$

and a_0 is determined as the remainder.
If the quotient is now divided by r, a_1 is the second remainder.
If the second quotient is divided by r, a_2 is the third remainder.
By successive division in this way, we find the successive remainders which are the values of the required digits

$$a_0, a_1, a_2, \ldots, a_n$$

EXAMPLES 2.2

(A) Express the decimal number 6317 in octal notation.

8)6317
8) 789 + 5
8) 98 + 5
8) 12 + 2
8) 1 + 4
0 + 1

Thus 6317 (decimal) = 14255 (octal)

(B) Express 247 decimal in binary.

2)247
2)123 + 1
2) 61 + 1
2) 30 + 1
2) 15 + 0
2) 7 + 1
2) 3 + 1
2) 1 + 1
2) 0 + 1

Thus 247 (decimal) = 11110111 (binary)

Binary numbers are most easily converted to decimal by reversing the above process. We check the last example by converting the binary number 11110111 to decimal by multiplying the left hand digit by 2, adding in the next digit and continuing this process, thus

1	1	1	1	0	1	1	1
	3	7	15	30	61	123	247

This conversion may, of course, be performed by successive division by 10 (decimal) thus

10)11110111
10) 11000 + 7
10) 10 + 4
0 + 2

but this method is very inconvenient.

(C) A number expressed in any base may be converted to decimal by the multiplication process of the last example. Thus 3215 (base 6) is converted to 731 (decimal) as follows:

3   2   1   5
$(6 \times 3 + 2 = 20)$ $(6 \times 20 + 1 = 121)$ $(6 \times 121 + 5 = 731)$

(D) Express the number 1660532 in the scale of seven in base 12 (T).

```
T)1660532
T) 110435 + 5
T)   4502 + 2
T)    251 + e
T)     14 + 2
        0 + e
```

Number to base T is $e2e25$.

(E) By converting the binary number 1111110 to decimal, find its binary factors.

1	1	1	1	1	1	0
	3	7	15	31	63	126

The factors of 126 are 2, 7 and 9.
These factors in binary are 10, 111, 1001 respectively. It may be verified that they are the factors of 1111110 by direct multiplication thus

```
 1001
  111
 1001
 1001
1001
111111
```

and $111111 \times 10 = 1111110$

2.3 Fractions

In decimal arithmetic,

$$9.367 \text{ means } 9 + \frac{3}{10} + \frac{6}{10^2} + \frac{7}{10^3}$$

Similarly in binary arithmetic,

$$1.101 \text{ means } 1 + \frac{1}{2} + \frac{0}{2^2} + \frac{1}{2^3}$$

In the scale of r,

$$2.512 \text{ means } 2 + \frac{5}{r} + \frac{1}{r^2} + \frac{2}{r^3}$$

As in the last example, fractions in the scale of r may be expressed in a form analogous to ordinary decimal fractions. In general, such a fraction will be equivalent to

$$\frac{b_1}{r} + \frac{b_2}{r^2} + \frac{b_3}{r^3} + \cdots$$

where b_1, b_2, b_3, ... are integers, each less than r, and one or more may be zero.

To express a given fraction F in the scale of r, it will be necessary to find digits b_1, b_2, b_3, ... such that

$$F = \frac{b_1}{r} + \frac{b_2}{r^2} + \frac{b_3}{r^3} + \cdots$$

Multiplying by r, we have

$$Fr = b_1 + \frac{b_2}{r} + \frac{b_3}{r^2} + \cdots$$

so that b_1 is the integral part of Fr whilst the fractional part F_1 is given by

$$F_1 = \frac{b_2}{r} + \frac{b_3}{r^2} + \cdots$$

On again multiplying by r, we have

$$F_1 r = b_2 + \frac{b_3}{r} + \cdots$$

so that b_2 is the integral part of $F_1 r$ whilst the fractional part F_2 is given by

$$F_2 = \frac{b_3}{r} + \cdots$$

Successive multiplications by r will give the values of b_3, b_4, ... If, at any stage, multiplication by r yields a product which is an integer, the process terminates and the fraction is expressible by a finite number of digits.

Otherwise, the process will not terminate in which case the digits will recur (analogous to a recurring decimal).

To PROVE that in the scale of r, the sum of the digits of any integer when divided by $(r - 1)$, gives the same remainder as when the integer itself is divided by $(r - 1)$.

Let N be the integer having digits $a_n, a_{n-1}, \ldots, a_1, a_0$ in that order so that

$$N = a_n r^n + a_{n-1} r^{n-1} + \cdots + a_1 r + a_0$$

Let S be the sum of the digits of N so that

$$S = a_n + a_{n-1} + \cdots + a_1 + a_0$$

Therefore

$$N - S = a_n(r^n - 1) + a_{n-1}(r^{n-1} - 1) + \cdots + a_1(r - 1)$$

Every term on the right-hand side of the above equation is divisible by $(r - 1)$. Therefore

$$\frac{N}{r - 1} - \frac{S}{r - 1} = \text{An integer}$$

and the proposition follows.

COROLLARY: In the decimal scale $(r = 10)$, a number is divisible by 9 if the sum of its digits is divisible by 9.

EXAMPLES 2.3

(A) Express $3/10$ in the scale of seven.

$$\frac{3}{10} \times 7 = 2 + \frac{1}{10}$$

$$\frac{1}{10} \times 7 = 0 + \frac{7}{10}$$

$$\frac{7}{10} \times 7 = 4 + \frac{9}{10}$$

$$\frac{9}{10} \times 7 = 6 + \frac{3}{10}$$

$$\frac{3}{10} \times 7 = 2 + \frac{1}{10} \quad \text{and the digit 2 recurs}$$

Therefore $3/10$ become $0.\dot{2}04\dot{6}$ in the scale of seven.

(B) Convert 2017.15625 decimal to the scale of twelve.

The integral and fractional parts will be converted separately.

```
12)2017
12) 168 + 1
12)  14 + 0
12)   1 + 2
      0 + 1
```

2017 decimal is 1201 in the scale of 12

```
   0.15625
  12
  (1).87500
  12
  (10).500
  12
  (6).
```

0.15625 decimal is 0.1*t*6 in the scale of 12

Required number is 1201.1*t*6

BINARY ARITHMETIC

The rules governing the use of the decimal point apply also to the use of the binary point. For example,

$$\text{binary number } 1.011 = 1 + \frac{0}{2} + \frac{1}{2^2} + \frac{1}{2^3}$$

$$= \frac{2^3 + 2 + 1}{2^3} = \frac{1011}{1000}$$

Again multiplication or division by a power of 2 involves moving the binary point, just as multiplication or division by a power of 10 is carried out by moving the decimal point. For example

$$101.1101 \times 1000 = 101110.1$$
$$1100101 \div 100 \; = \; 11001.01$$

Referring to Example 2.1F it follows that

$$110.101 \times 1.011 = \frac{110101 \times 1011}{1000 \times 1000}$$

$$= 1001.000111$$

This example may be conveniently set out thus:

```
 110.101
   1.011
 110.101
   1.10101
   0.110101
1001.000111
```

A division may be carried out thus:

$$\frac{10001.111}{11.01} = \frac{10001.111}{11.01} \times \frac{100}{100} = \frac{1000111.1}{1101}$$

```
1101)1000111.1(101.1
     1101
      10011
       1101
        1101
        1101
```

EXAMPLE 2.4
Express (a) 11/16, (b) 1/12 as binary fractions

(a) $\dfrac{11}{16} \times 2 = 1 + \dfrac{6}{16} = 1 + \dfrac{3}{8}$

$\dfrac{3}{8} \times 2 = 0 + \dfrac{3}{4}$

$\dfrac{3}{4} \times 2 = 1 + \dfrac{1}{2}$

$\dfrac{1}{2} \times 2 = 1 + 0.$

Thus 11/16 (decimal) = 0.1011 (binary)

(b) $\dfrac{1}{12} \times 2 = 0 + \dfrac{1}{6}$

$\dfrac{1}{6} \times 2 = 0 + \dfrac{1}{3}$

$\dfrac{1}{3} \times 2 = 0 + \dfrac{2}{3}$

$\dfrac{2}{3} \times 2 = 1 + \dfrac{1}{3}$

$\dfrac{1}{3} \times 2 = 0 + \dfrac{2}{3}$ so that the digits 0 and 1 will recur.

$\dfrac{1}{12}$ (decimal) = $0.000\dot{1}$ (binary)

EXERCISES 2.1

1. Add 10011011 to 111010 (binary). Subtract the latter number from the former. Check by converting the numbers to the decimal system. [11010101; 1100001; 155, 58, 213, 97]

2. Add $8ee5$, $14tt$, $730e$ in the scale of 12 and subtract the smallest from the largest. Check by converting each number to the decimal scale. [157e2; 7707; in the decimal scale, the three numbers are respectively 15545, 2434 and 12539]

3. Find (i) 273.561 × 34.21 in octal   [12272.02601]
   (ii) $940.e6t \div 7.3$ in duodecimal [135.61; rem. 0.57]

4. Find (i) 1101.011 × 110.101 (binary) [1011000.100111]
   (ii) 100000.1 ÷ 1.01 (binary) [11010]
   Check by converting to decimal.
   [(i) $13\frac{3}{8}$; $6\frac{5}{8}$; product $88\frac{39}{64}$][(ii) $32\frac{1}{2}$; $1\frac{1}{4}$: quotient 26]

5. Evaluate as binary fractions to 5 significant figures:

   (i) $\dfrac{11}{101}$ (binary) (ii) $\dfrac{10.1}{110.01}$ (binary) (iii) $\dfrac{10011}{11010}$ (binary)

   [(i) $0.\dot{1}00\dot{1}$ (ii) $0.01\dot{1}0\dot{0}$ (iii) 0.10111]

6. Convert 1552/2626 in the scale of seven to a fraction in the scale of ten. [$\frac{5}{8}$]

7. Convert (i) 4891 (decimal) to the scale of four
   (ii) 60435 (decimal) to octal
   (iii) 10087 (decimal) to duodecimal (scale of twelve)
   (iv) 200.211 in the scale of three to the scale of nine
   (v) 235.812 in the scale of nine to the scale of five.

[(i) 1030123   (ii) 166023   (iii) 5*t*07   (iv) 20.73   (v) 1234.4224 . . .]

8. Express (*a*) $\frac{7}{8}$ (*b*) $\frac{11}{16}$ (*c*) $\frac{1}{3}$ (*d*) $\frac{1}{12}$ (all decimal) as binary fractions.
   [(*a*) 0.111   (*b*) 0.1101   (*c*) 0.0\dot{1}   (*d*) 0.000\dot{1}]

9. In what scale is 25/128 (decimal) denoted by 0.0302? [four]

10. The order of the digits of an integral number *n* in the scale of *r* is altered in any way to form an integral number *m*. Show that $n \sim m$ is divisible by $(r - 1)$. (*Note:* $n \sim m$ means the difference of *n* and *m*.)

3 Digital computers—flow diagrams and the elements of programming

3.1 Historical Introduction

The first computing machines made in the United Kingdom were those invented by Babbage in the early part of the 19th century. They were known as the Difference Engine (1812–22) and the Analytical Engine (started 1832). In Babbage's Analytical Engine, provision was made for the five essential constituent parts of a modern computer, namely—the store, arithmetic unit, control unit, input and output devices.

The slowness of mechanical computers and their limited storage capacity led inevitably to the adoption of computers employing electronic devices for calculation, control and storage.

In 1946, a report on computers by von Neumann advocated

(a) The adoption of electronic principles

(b) The coding and storage of operating instructions in digital form

(c) The storage of numbers and instructions as "words" in a common store, words being represented by trains of electrical pulses

(d) The use of the binary scale

(e) Decimal to binary conversion at input, and binary to decimal at output.

Prototypes of electronic digital computers were developed by the Universities of Cambridge and Manchester (1948–50) and amongst the earliest digital computers built commercially was Ferranti I produced by Ferranti Ltd. in 1951.

3.2 Computer Programming

The primary function of an electronic computer, like that of a desk calculating machine, is to perform numerical computations quickly and accurately. In a desk machine, the numbers are set up by keys and the arithmetical operations involved are performed by pressing buttons. Answers are displayed or copied on paper.

In an electronic computer, the arithmetical operations are performed by the *arithmetic unit*. If it were necessary to set up the numbers manually and copy down the results of calculations, the inherent speed of the electronic computer would be wasted. Instead numbers are stored in electrical form within the machine and methods of obtaining rapid input and output of data have been devised. Instead of pressing buttons to initiate arithmetic operations, instructions are written into the input to inform the arithmetic unit how to manipulate the numbers in the stores.

These instructions are called a *program* and are stored within the computer in precisely the same form as numbers are stored.

3.3 The Essential Features of Electronic Computers

The essential constituents of an electronic digital computer are
- (*a*) Input and output devices
- (*b*) Storage for numbers and instructions
- (*c*) Arithmetic unit
- (*d*) Control unit.

Digital computers work in the binary scale rather than the decimal. This is because it is simplest to employ electronic devices in two states, "on" and "off", representing 1 and 0 respectively, rather than in ten.

(a) INPUT AND OUTPUT SYSTEMS

The most widely used forms of input comprise
punched paper tape
punched cards
magnetic tape.
Inputs in these forms are prepared away from the computer by peripheral or "off-line" equipment, the input data being represented in coded form.

Paper tapes are prepared by teletypes which punch holes across the width of the tape to represent numbers or other characters in binary number form in accordance with a code which the computer in use is able to read. Tapes punched in 5, 6, 7 and 8 channels are in use. The pattern of holes can be regarded as representing binary numbers, each space and hole representing the binary digits* 0 and 1 respectively. An 8-channel tape provides space for a

* The term "binary digits" is usually abbreviated to "bits".

TABLE 3.1

Teletype symbol	Appearance of the tape							
A	o					o		o
B	o					o	o	
C	o	o				o	o	o
D	o				o	o		
E	o	o			o	o		o
F	o	o			o	o	o	
G	o				o	o	o	o
H	o			o	o			
I	o	o		o	o			o
J	o	o		o	o		o	
K	o			o	o		o	o
L	o	o		o	o	o		
M	o			o	o	o		o
N	o			o	o	o	o	
O	o	o		o	o	o	o	o
P	o		o	o				
Q	o	o	o	o				o
R	o	o	o	o			o	
S	o		o	o			o	o
T	o	o	o	o		o		
U	o		o	o		o		o
V	o		o	o		o	o	
W	o	o	o	o		o	o	o
X	o	o	o	o	o			
Y	o		o	o	o			o
Z	o		o	o	o	o		
space	o	o		o				
0		o	o	o				
1	o	o	o	o				o
2	o	o	o	o		o		
3		o	o	o			o	o
4	o	o	o	o	o			
5		o	o	o	o			o
6		o	o	o	o	o		
7	o	o	o	o	o	o	o	
8	o	o	o	o	o			
9		o	o	o	o			o
,	o	o		o	o	o		
&		o		o	o	o	o	
;	o	o	o	o	o		o	o
:		o	o	o	o		o	
—		o		o	o	o		o
+		o		o	o		o	o
=	o	o	o	o	o			o

maximum of 8 holes across the width of the tape so that $2^8 = 256$ different combinations of spaces and holes are possible, representing the numbers 0–255 in binary form. In practice, not all the possibilities are employed for coding.

Table 3.1 shows that part of an 8-channel paper tape code which deals with the coding of the alphabet, the decimal digits and a few other characters. (The small holes are sprocket holes.) This code is now widely used for punching programs written for modern computers. It will be noted that each coding contains an even number of holes, each representing binary 1. If, due to error, the equipment punches one hole too few or too many, the resulting odd number of holes can be detected by the computer. In such cases, an error message will be output by the computer so that the appropriate coding may be corrected.

Paper tape is usually read electrically by using a light and a photoelectric cell, but sometimes mechanically by a metal probe which completes a circuit through any hole with a metal plate. In both cases, each hole gives rise to an electric pulse. A program is thus converted into a train of pulses capable of operating the computer.

A computer is frequently employed to deal with information which has to be stored and which may have to be amended from time to time. One convenient way of handling data of this kind is to use *punched cards* as the input medium. Figure 3.1 illustrates such a card.

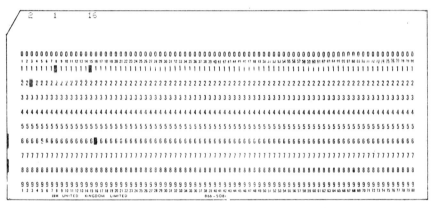

FIGURE 3.1   A punched card.

Each hole or combination of holes in a column of a punched card represents an item of information, which is read into the computer as a train of pulses, by means of a photoelectric cell or a metal brush.

Magnetic tapes carrying numerical and other information in the coded form of binary numbers may also be used as input devices. A tape usually consists of a thin layer of magnetic material with a superimposed layer of plastic, or the magnetic material is sometimes placed between two layers of plastic. Tapes are usually 1/2 in. wide and carry 6 to 12 channels. Across the width of such a tape, a binary number may be represented by the state of magnetization of from 6 to 12 elements of magnetic material, one in each channel. The binary digit 1 corresponds to a particular direction of magnetization of an element, and 0 to the opposite direction.

Punched cards, line printers and paper and magnetic tapes are the most widely used output systems.

(b) STORAGE

Modern computers usually incorporate magnetic core storage employing, of course, the two states of magnetization. This type of storage usually holds the program and data to be used immediately in carrying out the program. Other forms of storage are used to provide *backing storage*, such as magnetic tapes, discs and drums, punched cards and paper tapes. The choice of the method of storage depends upon the nature of the work for which the computer is used, e.g. when processing business data, for which a large storage capacity is required, magnetic discs are almost always used.

(c) ARITHMETIC UNIT

This is the part of the computer system in which arithmetic and logical operations are performed upon special storage locations called *registers*.

(d) CONTROL UNIT

The storage, arithmetic, input and output units of a computer are interconnected through groups of switches called *gates*. These gates are opened and closed by control wave forms which emanate from the control unit. This unit ensures that these signal control pulses arrive at the precise moment required (as referred to "clock pulses" generated by a crystal oscillator or an equivalent device) to allow certain operations to be carried out upon the subsequent arrival of the appropriate coded instruction.

Figure 3.2 is a schematic diagram showing the essential interconnections between the basic computer units.

3.4 Words and Addresses

The central processor of a computer contains a store which is usually some form of magnetic core storage. The central store, and the other types of storage within a particular computer, store the data and instructions required

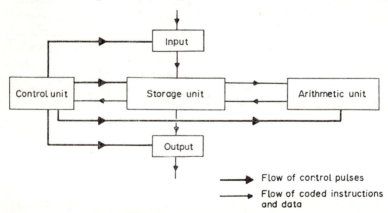

FIGURE 3.2 *Interconnections of computer units.*

for the computation and also hold other information as required at various stages of the computation. Both characters and numbers are stored in binary coded form.

Words are ordered sets of binary digits representing instructions or quantities held in storage compartments called *locations*. The length of a word is specified by the number of binary digits (i.e. bits) which it contains.

The *address* of a word is its reference—usually a number. The computer must be given the reference to enable it to find a particular item of data stored in the word. All data in storage must be addressable.

3.5 Flow Diagrams and Programs

A *flow diagram* is an exposition in diagrammatic form of the logical sequence of steps which must be followed to carry out a computation, taking into account every possible contingency.

A *program* is an interpretation of the flow diagram as a body of instructions for carrying out the actual arithmetical work, and for controlling the order in which the arithmetic instructions are to be obeyed. A program devised to perform a given computation is not necessarily unique. It must be the aim of the programmer to produce the program which will carry out the computation in the most simple and expeditious manner. The necessary instructions are fed into the computer in coded form, e.g. in Autocode, Algol or Fortran.

The following symbols are widely used in flow diagrams:

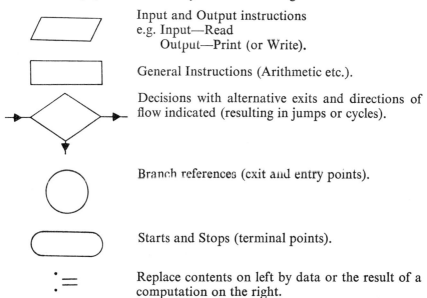

Input and Output instructions
e.g. Input—Read
Output—Print (or Write).

General Instructions (Arithmetic etc.).

Decisions with alternative exits and directions of flow indicated (resulting in jumps or cycles).

Branch references (exit and entry points).

Starts and Stops (terminal points).

Replace contents on left by data or the result of a computation on the right.

The addresses of various stores will be denoted by capital letters A, B, C, ... (Autocodes) or by the use of names (Algol and Fortran)

For simplicity, "the store whose address is A" will be called "Store A".

The instruction $\boxed{R := \dfrac{B}{C}}$

requires the computer to divide the number in Store B by the number in Store C and place the result in Store R.

The instruction $\boxed{I := I + 1}$

is a counting instruction to ensure that a part of the computation which is to be repeated (a cycle) is carried out an appropriate number of times. If the number in Store I is initially 1, I will successively hold the numbers 1, 2, 3, ... until the computer is instructed to terminate the cycle.

A computer is capable of comparing two numbers. An instruction in the flow diagram, inviting such a comparison, is usually written within a diamond-shaped decision box. For example, the computer may be asked "Is $D > 0$?", i.e. is the number in Store D greater than zero?

The symbol   No 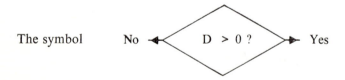 Yes

demands one of the two answers, "Yes" or "No", corresponding to the two outputs from the box, resulting in one or other of two courses of action.

EXAMPLE 3.1

Solve $ax^2 + bx + c = 0$.

$$x = -\frac{b}{2a} \pm \frac{\sqrt{(b^2 - 4ac)}}{2a}$$

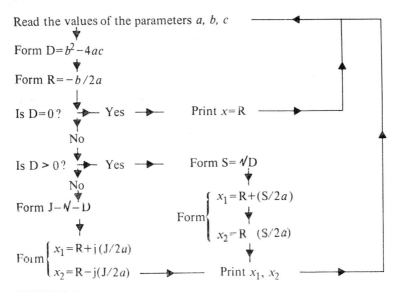

Read the values of the parameters a, b, c

Form $D = b^2 - 4ac$

Form $R = -b/2a$

Is $D = 0$? $\longrightarrow$ Yes $\longrightarrow$ Print $x = R$

No

Is $D > 0$? $\longrightarrow$ Yes $\longrightarrow$ Form $S = \sqrt{D}$

No

Form $J - \sqrt{-D}$

Form $\begin{cases} x_1 = R + (S/2a) \\ x_2 = R \ (S/2a) \end{cases}$

Form $\begin{cases} x_1 = R + j(J/2a) \\ x_2 = R - j(J/2a) \end{cases}$ $\longrightarrow$ Print x_1, x_2

FIGURE 3.3

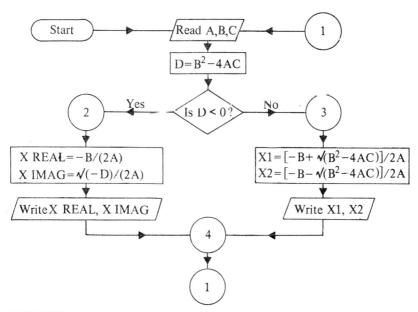

Start $\longrightarrow$ Read A,B,C $\longleftarrow$ 1

$D = B^2 - 4AC$

2 $\longleftarrow$ Yes — Is $D < 0$? — No $\longrightarrow$ 3

X REAL $= -B/(2A)$
X IMAG $= \sqrt{(-D)}/(2A)$

$X1 = [-B + \sqrt{(B^2 - 4AC)}]/2A$
$X2 = [-B - \sqrt{(B^2 - 4AC)}]/2A$

Write X REAL, X IMAG

Write X1, X2

4

1

FIGURE 3.4

The problem must be analysed to devise a concise logical plan to solve the equation whether the roots be real or complex. Figure 3.3 is a suitable flow diagram.

Using the symbols of Section 3.5, with A, B, C, D, X1, X2, X REAL, X IMAG as names of store addresses, the flow diagram may be written more concisely as in Fig. 3.4.

The program for this computation in Fortran would appear as follows:

1 READ A, B, C   The equation parameters are read.

D = B*B — 4.0*A*C   Discriminant, D, evaluated. Ex-
IF (D) 2, 3, 3   amination of D decides the type
                                         of solution. For complex roots
                                         the program jumps to label 2.
                                         For D = 0, D > 0 the program
                                         jumps to label 3.

3 X1 = (−B + SQRT (D))/(2.0*A)   Real roots formed; SQRT calls
   X2 = (−B − SQRT (D))/(2.0*A)   up function which forms square
                                         roots.

WRITE X1,X2   Read roots output.
GO TO 4   Jump to label 4 to avoid that
                                         part of the program which deals
                                         with complex roots.

2 X REAL = −B/(2.0*A)
   X IMAG = SQRT (−D)/(2.0*A)   Solution for complex roots.
   WRITE X REAL, X IMAG

4 GO TO 1   The program returns to read
                                         data for another set of param-
                                         eters.

For simplicity, information concerning the manner in which the program is terminated and the use of formats for input and output have been omitted. Certain directives, such as MASTER and END, have also been omitted.
NOTE: * is the symbol for multiplication in Algol and Fortran languages.

EXAMPLE 3.2

Calculate the sum of the terms on the leading diagonal of an $(N \times N)$ matrix.

$$\begin{pmatrix} A_0 & A_1 & A_2 & A_3 & \cdots & A_{N-1} \\ A_N & A_{N+1} & A_{N+2} & A_{N+3} & \cdots & A_{2N-1} \\ A_{2N} & A_{2N+1} & A_{2N+2} & A_{2N+3} & \cdots & A_{3N-1} \\ A_{3N} & \cdots & \cdots & A_{3N+3} & \cdots & \cdots \\ \vdots & & & & & \vdots \\ A_{(N-1)N} & \cdots & \cdots & \cdots & \cdots & A_{N^2-1} \end{pmatrix}$$

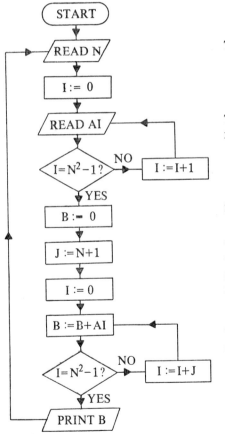

The order of the matrix is stored.

The matrix is read and stored by means of a repeating cycle or loop.

Store for running total set to zero.

The numbers of the elements to be summed increase in arithmetic progression by $(N + 1)$ starting at 0 and finishing at $N^2 - 1 = (N - 1)(N + 1)$.

This cycle adds the appropriate A elements into the accumulator B.

This jump is used to deal with the data of a new matrix.

FIGURE 3.5

EXAMPLE 3.3
Evaluate π using the series

$$\frac{\pi^4}{96} = \frac{1}{1^4} + \frac{1}{3^4} + \frac{1}{5^4} + \cdots$$

ignoring terms less than 10^{-8}.

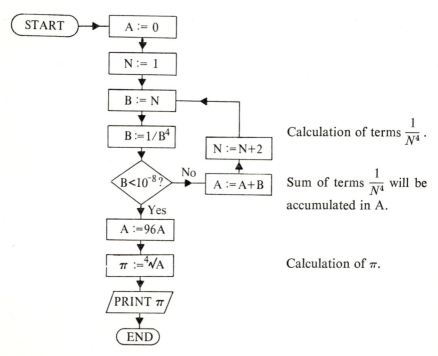

Calculation of terms $\dfrac{1}{N^4}$.

Sum of terms $\dfrac{1}{N^4}$ will be accumulated in A.

Calculation of π.

FIGURE 3.6

EXAMPLE 3.4
Compute the mean, variance and standard deviation of frequency weighted values of a variate x (frequency f).

(There is an increasing tendency to write flow diagrams and programs in plain language. Algol and Fortran languages reflect this tendency, and the following flow diagram and Algol program illustrate this development.)

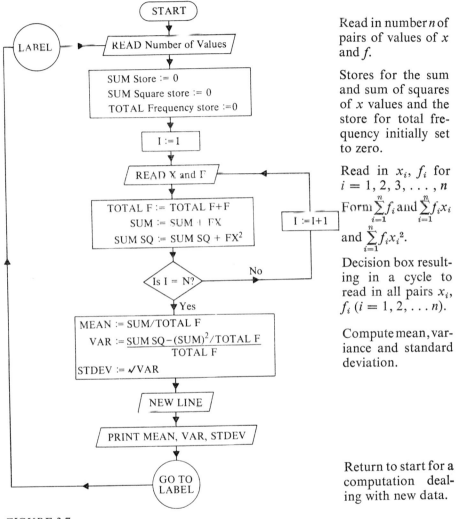

START

LABEL

READ Number of Values

SUM Store := 0
SUM Square store := 0
TOTAL Frequency store := 0

I := 1

READ X and F

TOTAL F := TOTAL F+F
SUM := SUM + FX
SUM SQ := SUM SQ + FX^2

I := I+1

Is I = N?   No

Yes

MEAN := SUM/TOTAL F
VAR := $\dfrac{\text{SUM SQ} - (\text{SUM})^2/\text{TOTAL F}}{\text{TOTAL F}}$
STDEV := √VAR

NEW LINE

PRINT MEAN, VAR, STDEV

GO TO LABEL

Read in number n of pairs of values of x and f.

Stores for the sum and sum of squares of x values and the store for total frequency initially set to zero.

Read in x_i, f_i for $i = 1, 2, 3, \ldots, n$

Form $\sum_{i=1}^{n} f_i$ and $\sum_{i=1}^{n} f_i x_i$ and $\sum_{i=1}^{n} f_i x_i^2$.

Decision box resulting in a cycle to read in all pairs x_i, f_i ($i = 1, 2, \ldots n$).

Compute mean, variance and standard deviation.

Return to start for a computation dealing with new data.

FIGURE 3.7

ALGOL PROGRAM

Below is an Algol program to compute the mean, variance and standard deviation of frequency weighted values of a variate X.

```
BEGIN INTEGER I, N, F, TOTALF:
      REAL SUM, SUMSQ, MEAN, VAR, STDEV, X;
      SWITCH S := START;

      START:READ N;
            SUM := SUMSQ := 0;                    TOTALF := 0;
            FOR I:=1 STEP 1 UNTIL N DO
                  BEGIN READ X,F;
                        TOTALF:=TOTALF+F;
                        SUM:=SUM+F*X;
                        SUMSQ:=SUMSQ+F*X*X;

                  END;
            MEAN:=SUM/TOTALF;
            VAR:=(SUMSQ-(SUM*SUM)/TOTALF)/TOTALF;
            STDEV:=SQRT(VAR);

            PRINT MEAN,VAR,STDEV;
            GO TO START;

END;
```

EXERCISES 3.1

Write flow diagrams for the following computations:

1. Read three numbers and arrange them in numerical order, placing the smallest in store position Y_1 and the largest in Y_3. Extend the method to deal with the general case where N numbers are read.

2. Calculate the income tax to be paid by a person earning £S per annum who receives allowances of £A if
   (i) the lowest rate of tax is X pence per £ for the next £B.
   (ii) the middle rate of tax is Y pence per £ for the next £C.
   (iii) the highest rate of tax is Z pence per £ for the remainder.
   (The values of A, B, C, S, X, Y, Z are to be read.)

3. Solve
   $$a_1x + b_1y = c_1$$
   $$a_2x + b_2y = c_2.$$

4. Group n integers between 0 and 99 into classes
   $$0–9, 10–19, 20–29, \ldots, 90–99$$
   and print out the number of integers in each class.

5. Evaluate π using the formula
$$\frac{2}{\pi} = \frac{\sqrt{2}}{2} \times \frac{\sqrt{(2 + \sqrt{2})}}{2} \times \frac{\sqrt{\{2 + \sqrt{(2 + \sqrt{2})}\}}}{2} \times \cdots.$$

until factors differ from unity by less than 10^{-6}.

6. Compute and print the prime numbers lying between 10 and 1 000 000 by dividing each successive odd number in this range by all odd numbers up to its square root and testing for a remainder.

7. Evaluate the product
$$\prod_{p=2}^{n} \left(1 - \frac{(-1)^p}{p^2}\right)$$

8. Read angle A in degrees. Replace it by the related angle $B°$ in the first quadrant. Convert $B°$ into X radians and compute $\cos(X)$ to 6 decimal places accuracy using the series
$$\cos(X) = 1 - \frac{X^2}{2!} + \frac{X^4}{4!} - \frac{X^6}{6!} + \cdots$$

Calculate the distance D from the earth of a satelite revolving around the earth in an elliptic orbit from the relationship
$$D = K/(1 + E \cos A)$$

The angle A and the constants K and E are to be read.

9. Fit a straight line $y = a + bx$ to n points (x_i, y_i) by the method of least squares. The equations for a and b are
$$\sum y_i = na + b \sum x_i$$
$$\sum x_i y_i = a \sum x_i + b \sum x_i^2$$

4 Sets

4.1 The Idea of a Set

The term "set", along with many other terms which are equivalent to it, are in everyday use. For example, we speak of a set of dishes, a set of golf clubs, a set of games in tennis, a pack of cards, a flock of sheep, a herd of cattle and so on. Intuitively, we think of a set as a collection of objects.

The idea of a set is a very important mathematical concept and one might, without exaggeration, describe mathematics as the study of the properties of various kinds of sets.

In arithmetic we work with a set of numbers, whilst in geometry we are concerned with sets of points, sets of lines, sets of triangles, and so on. The term "set of real numbers" has already been informally introduced in Section 1.8.

The individual members of a set are called its *elements* which may be finite or infinite in number, e.g. the set of cards in a pack comprises 52 elements whilst the set of positive integers has an infinite number of elements.

A set is defined if all its elements are known. Thus a set will be well defined if its elements are listed or specified by a rule or a description. A *well-defined* set is a set associated with a property by which it may be decided whether any object is or is not an element of the set.

Examples of well-defined sets are

(i) The set of all counties in the United Kingdom. The elements of this set may be listed, viz. {Aberdeenshire, Anglesey, . . . , Worcestershire, Yorkshire}

(ii) The set of all symphonies composed by Beethoven, comprising a list of nine

(iii) The set of all positive even integers less than 16. Its elements may be listed as {2, 4, 6, 8, 10, 12, 14}

(iv) The set of all points on the circumference of a given circle. Its elements are infinite in number and cannot be listed.

The collection of all living persons who will eventually become Members of Parliament is an example of a set which is not well defined, since clearly it is impossible to decide whether any particular person is a member of the set.

In future, the term "set" will be assumed to mean a well-defined set.

4.2 Notation

We shall use capital letters $A, B, C, \ldots$ to denote sets and small letters $a, b, c, \ldots$ to denote members or elements of sets.

$x \in A$ means that x is an element of the set A and is frequently read as "x belongs to A" or "x is a member of A". The negative of the statement $x \in A$ is written $x \notin A$ and may be read as "x does not belong to A".

For example, if A is the set of odd integers

$$3 \in A \qquad \text{but} \qquad 4 \notin A$$

Let $x_1, x_2, \ldots x_n$ be the elements of the set A. We may represent the set A by listing all its elements and enclosing them within brackets $\{ \ \}$ for which we read "the set of". Thus we may write

$$A = \{x_1, x_2, \ldots x_n\}$$

The symbol : is read as "such that" so that the set A may be represented more briefly as

$$A = \{x_i : i = 1, 2, \ldots n\}$$

or even

$$A = \{x_i\}$$

In Section 4.1, the set (iii) may be written as either

$$\{2, 4, 6, 8, 10, 12, 14\}$$

or

$$\{\text{all positive even integers less than } 16\}.$$

The elements of the set (iv) in Section 4.1 cannot be listed but may be written

$$\{\text{points on the circumference of a given circle}\}$$

If A is a set of objects x satisfying a property P, we may write

$$A = \{x:P\}$$

For example, if A is the set of all integers whose squares are greater than or equal to 25, we may write

$$A = \{x : x \text{ is an integer and } x^2 \geqslant 25\}$$

This is a convenient brief description of a set which has an infinite number of elements.

Certain phrases occur so frequently as to justify the use of notation for them. For the phrase, "There exist(s)", we shall write $\exists$.

If p and q represent statements, the phrase "p implies q" or "if p then q" is written $p \Rightarrow q$ which also means that p is a sufficient condition for q and also that q is a necessary condition for p. If p and q are equivalent statements, i.e. $p \Rightarrow q$ and $q \Rightarrow p$, we write $p \Leftrightarrow q$ which may be read as "p if and only if q." For example, if p is the statement that $x^2 + 12 = 7x$ and q is the statement that x is either 3 or 4, then

$$x^2 + 12 = 7x \Leftrightarrow x = 3 \text{ or } x = 4$$

EQUALITY OF SETS

Two sets A and B are equal, i.e. $A = B$, means that the two sets have precisely the same elements. Thus, if x is an element of A, x is also an element of B and vice versa.

$$x \in A \Rightarrow x \in B$$
$$x \in B \Rightarrow x \in A$$

or

$$x \in A \Leftrightarrow x \in B$$

ONE-TO-ONE CORRESPONDENCE

Each car on the roads of Britain must display its own particular registration number by which it may be identified, i.e. one registration number corresponds to one car, and vice versa. There is said to be a one-to-one (1–1) correspondence between cars and their registration numbers.

There are many other instances of 1–1 correspondences, e.g.
  (i) Each voter and his name on the Electoral Register
 (ii) Each Member of Parliament and his constituency
(iii) Each seat in a classroom and the student occupying it (it is assumed that all seats are occupied)
 (iv) Each £1 note and its number.

The following are examples of many-to-one and one-to-many correspondences:
  (i) Sons and daughters and their mother
  (ii) Ships and their passengers.

EQUIVALENT SETS

Two sets are said to be *equivalent* if there is a one-to-one correspondence between their elements. The equivalence of two sets A and B is denoted by $A \sim B$.
  The following are examples of equivalent sets:
  (i) {Husbands} $\sim$ {Their wives} assuming monogamy and that the two sets contain no widowers and no widows.
  (ii) {National flags} $\sim$ {Corresponding nations}
  (iii) {Thumbprints} $\sim$ {Persons whose prints they are}
  (iv) $\{1, 2, 3, \ldots, 10\} \sim$ {Fingers and thumbs on a pair of hands}

SUBSETS

Let A and B be sets. A is a *subset* of B provided that every element of A is also an element of B. We denote that A is a subset of B by writing $A \subset B$ or $B \supset A$. The notation $A \subset B$ may be read as "A is included in B". The inclusion of A in B may also be expressed symbolically thus

$$A \subset B \Leftrightarrow (x \in A \Rightarrow x \in B)$$

It follows, by definition, that any set is a subset of itself, i.e.

$$A \subset A$$

Again, if A and B are equal, A is a subset of B and B is a subset of A, and vice versa, i.e.

$$A = B \Leftrightarrow A \subset B \quad \text{and} \quad B \subset A$$

A method frequently employed to prove that two sets are equal is that of showing that each is a subset of the other.
  If every element of a set A is an element of a set B but B has one or more elements which are not in A, i.e.

$$A \subset B \quad \text{but} \quad B \neq A$$

A is said to be a *proper* subset of B.

Let $N = \{$positive integers$\} = \{1, 2, 3, 4, \ldots\}$
and let $E = \{$positive even integers$\} = \{2, 4, 6, 8, \ldots\}$
then

$E \subset N$ but $E \neq N$

and E is a proper subset of N.
  If $A = \{5, 7, 9, 11\}$, A is a proper subset of N.
  If $B = \{-3, 1, 2, 3, 4\}$, B is not a subset of N since B contains the element
-3 which is not an element of N. We write $B \not\subset N$.

EXERCISES 4.1

1. Describe the following sets:
    (i) {Beethoven, Mozart, Brahms, Tchaikowski, Sibelius, Dvorak}
    (ii) {Monday, Tuesday, Wednesday, Thursday, Friday, Saturday, Sunday}
    (iii) {triangles, quadrilaterals, pentagons, ...}
    (iv) $\{2, 4, 6, 8, \ldots\}$
    (v) $\{1, 2, 3, 5, 7, 11, 13, 17, \ldots\}$
    (vi) $\{5, 10, 15, 20, 25, \ldots\}$
    (vii) $\{7, 14, 21, 28, \ldots\}$
    (viii) $\{2, 2^2, 2^3, 2^4, \ldots\}$
    (ix) $\{123, 132, 213, 231, 312, 321\}$

2. List all the elements of the following sets and describe them by a suitable notation.
    (i) All positive integers whose squares are less than 36.
       $[\{1, 2, 3, 4, 5\} = \{x : x$ is a positive integer and $x^2 < 36.\}]$
    (ii) All positive integers whose squares are greater than 16 and less than 50.
    (iii) All positive integers whose square roots are less than 11.
    (iv) All prime numbers between 30 and 50.
    (v) All multiples of 3 from 0 to 99 inclusive.
    (vi) All two-digit numbers, each of whose unit digit is twice its tens digit. How
        many elements do the above sets contain?

3. Give an example of a set which may be described in more than one way.

4. Which of the following pairs of sets are equal?
    (i) $A = \{a, b, c, d\}$ $B = \{d, c, b, a\}$
    (ii) $C = \{$even numbers$\}$ $D = \{2, 4, 6, 8\}$
    (iii) $E = \{7, 11, 13, 17, 19\}$ $F = \{$prime numbers between 6 and 20$\}$

5. Invent sets which are equivalent to the following sets:
    (i) {pictures in an art exhibition}
    (ii) {houses in a street}
    (iii) {race horses}
    (iv) {major roads in Britain}
    (v) {sides of a triangle}
    (vi) {concentric circles}

6. If $A = \{a, b, c\}$ and $B = \{1, 2, 3\}$, show that there are six possible one-to-one
   correspondences between A and B.

7. If $A = \{1, 2, 3, 4\}$, construct all the possible subsets of the set A containing at least one element.

8. If $A = \{a, b, d, e\}$, $B = \{b, c, e, f\}$, $C = \{b, c\}$ and $D = \{a, b, c, d, e, f\}$, state which pairs of sets may be linked by the symbol $\subset$.

9. Use set language to describe the concepts of "point" and "line" in terms of each other.

4.3 Operations on Sets

We are familiar with the combination of numbers and algebraic quantities by the application of the operations of addition, subtraction, multiplication and division, e.g.

$$2 + 3 = 5 \qquad 2 \times 3 = 6$$

Let us now consider the operations which may be applied to combine sets.

UNION AND INTERSECTION

The *intersection* of two sets A and B, denoted by $A \cap B$, is the set of those elements which are members of both A and B, i.e.

$$x \in A \text{ and } x \in B \Leftrightarrow x \in A \cap B$$

$A \cap B$ is read "A intersection B" or "A cap B".

The *union* of two sets A and B, denoted by $A \cup B$, is the set of those elements which are members of either A or B or of *both* A and B, i.e.

$$x \in A \text{ or } x \in B \Leftrightarrow x \in A \cup B$$

$A \cup B$ is read "A union B" or "A cup B".

Let $A = \{1, 3, 5, 7\}$ and $B = \{2, 4, 5, 6, 7\}$. Then

$$A \cup B = \{1, 2, 3, 4, 5, 6, 7\}$$

$$A \cap B = \{5, 7\}$$

UNIVERSAL SET

The set of all possible objects to which we are referring in a given problem is called the *universal set* for that problem.

Thus if we wish to find the set of people living in a given town, whose ages exceed 50 years, the universal set might be the whole population of that town.

Again, for the set of all prime numbers between 0 and 1 000, the universal set might comprise all positive integers or it might be all positive integers from 1 to 1 000.

In plane geometry, the universal set might be the set of all points lying in a plane. It is clear that the universal set for a given problem is not necessarily unique.

We shall denote the universal set by V.

THE EMPTY SET $\varnothing$

The *empty* set, *null* set, or *void* set is the set which has no elements, and it is denoted by $\varnothing$. Although the idea of a set having no elements may appear, at first sight, to be a pointless abstraction, the concept is in fact useful and the symbol $\varnothing$ is needed.

If two sets A and B have no common elements, their intersection is the empty set, and we may write $A \cap B = \varnothing$. For example,

$$\{1, 3, 5\} \cap \{2, 4, 6, 8\} = \varnothing$$

Again, if A is the set of all boys in a class and B is the set of all girls,

$$A \cap B = \varnothing$$

APPLICATIONS

We shall illustrate these ideas by means of the following situation. In the VIth form of a school, there are 8 pupils who are studying mathematics. Let us call them a, b, c, d, e, f, g, h. The three subjects which each of these pupils study are tabulated below.

(Abbreviations: M = Mathematics, P = Physics, C = Chemistry, B = Biology, E = Economics.)

Pupil	Subjects studied			Pupil	Subjects studied		
a	M	P	C	e	M	E	B
b	M	P	C	f	M	E	C
c	M	P	C	g	M	E	B
d	M	E	C	h	M	P	C

The set Q of pupils who study C *and* E is $\{d, f\}$.

The set R of pupils who study C or E or *both* is $\{a, b, c, d, e, f, g, h\}$.

The set R is equal to the universal set V for this situation which may be taken to be the set of 8 pupils.

The set S of pupils who study E is $\{d, e, f, g\}$.

The set T of pupils who study C is $\{a, b, c, d, f, h\}$.

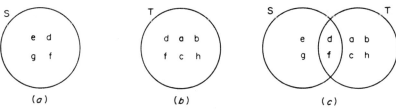

(*a*)   (*b*)   (*c*)

FIGURE 4.1

In Fig. 4.1(*a*), we represent the set S diagrammatically by a circle containing all the elements d, e, f, g of S. Similarly, in Fig. 4.1(*b*), we represent the set T by another circle enclosing all the elements a, b, c, d, f, h of T. From Fig. 4.1(*c*), we see that the sets S and T have the two elements d, f, and only these two elements, in common. The set $\{d, f\}$, i.e. the set Q is therefore the *intersection* of the sets S and T.

$$\{d, f\} = \{d, e, f, g\} \cap \{a, b, c, d, f, h\}$$

i.e.

$$Q = S \cap T$$

The set R comprises all the elements of the sets S and T together. Thus R is the *union* of S and T.

$$\{a, b, c, d, e, f, g, h\} = \{d, e, f, g\} \cup \{a, b, c, d, f, h\}$$

i.e.

$$R = S \cup T$$

For any two sets A and B, represented, as described above, by two circles, their intersection $A \cap B$ may be represented by the area common to the two circles and their union $A \cup B$ by the total area enclosed by the two circles as illustrated in Fig. 4.2. Diagrams such as these are called *Venn diagrams*.

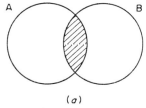   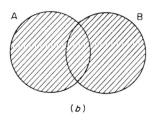

(*a*)   (*b*)

(*a*) *The shaded area*   (*b*) *The shaded area*
    represents $A \cap B$.   *represents* $A \cup B$.

FIGURE 4.2

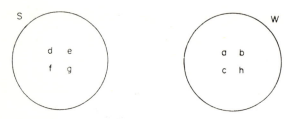

FIGURE 4.3

We now require what is the set of pupils who study both *P* and *E*? The table shows that this set has no members and is therefore the empty set ∅.

S = {Pupils who study E} = {d, e, f, g}

Let

W = {Pupils who study P} = {a, b, c, h}
$S \cup W$ = {d, e, f, g} ∪ {a, b, c, h} = {a, b, c, d, e, f, g, h}
        = V (the universal set)
$S \cap W$ = {d, e, f, g} ∩ {a, b, c, h} = ∅ (the empty set)

Two sets, like *S* and *W*, which have no common elements are said to be *disjoint* and their intersection is the empty set ∅. They may be represented by the Venn diagram of Fig. 4.3.

The set of pupils who do not study *E* is {a, b, c, h}. We see that it comprises all those elements of the universal set *V* which are *not* in *S* (this being the set of pupils who do study *E*). We denote the first set by *S'*.

S' is called the *complement* of *S* in *V*.

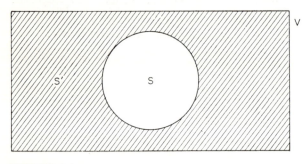

FIGURE 4.4

Formally, the complement S' of a set S is the set of all those elements of the universal set which are not elements of S.

$$S' = \{x : x \in V \text{ and } x \notin S\}$$

or

$$x \in V \quad \text{but} \quad x \notin S \Leftrightarrow x \in S'$$

The Venn diagram of Fig. 4.4 illustrates S and its complement S'. The universal set V is represented by the interior of a rectangle enclosing S and S'.

ASSOCIATIVE LAWS

The Venn diagrams for three sets A, B, C (Fig. 4.5) illustrate that

$$(A \cap B) \cap C = A \cap (B \cap C)$$

and

$$(A \cup B) \cup C = A \cup (B \cup C)$$

These results may be compared with the associative laws for addition and multiplication in ordinary algebra, namely,

$$(a + b) + c = a + (b + c)$$
$$(a \cdot b) \cdot c = a \cdot (b \cdot c)$$

It may therefore be said that sets obey the associative laws under the operations of union $\cup$ and intersection $\cap$ and that the brackets are not significant. In other words,

$(A \cap B) \cap C$ may be written without ambiguity, as $A \cap B \cap C$ and $(A \cup B) \cup C$ may be written $A \cup B \cup C$.

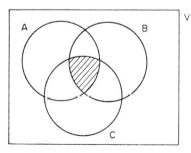

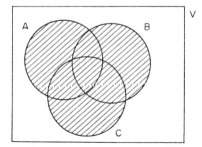

(a) *The shaded area represents* $(A \cap B) \cap C = A \cap (B \cap C)$.

(b) *The shaded area represents* $(A \cup B) \cup C = A \cup (B \cup C)$.

FIGURE 4.5

EXAMPLE 4.1

Draw Venn diagrams to show that

$$A \cup (B \cap C) = (A \cup B) \cap (A \cup C)$$

$A \cup (B \cap C)$ and $(A \cup B) \cap (A \cup C)$ are both represented by the same shaded area (Fig. 4.6).

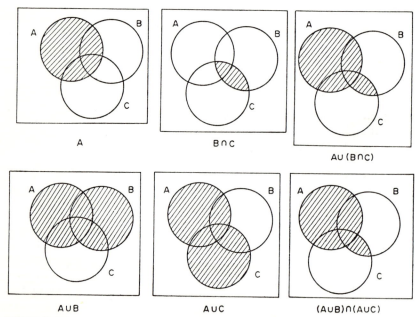

FIGURE 4.6

EXERCISES 4.2

1. By drawing Venn diagrams, illustrate the following:
   (a) $A \cap B = \emptyset$ (b) $A \cup B = V$ (c) $A \cup B$ and $A \cap B$ if $A \subset B$
   (d) $(A')'$ (e) $A \cap B \cap C = \emptyset$.

2. Simplify:
   (a) $A \cup A$ (b) $A \cap A$ (c) $A \cup \emptyset$ (d) $A \cap \emptyset$ (e) $A \cup V$ (f) $A \cap V$
   (g) $A \cup A'$ (h) $A \cap A'$

3. Simplify:
   (a) $A \cup (A \cap B)$ (b) $(A \cup B) \cap A$ (c) $(A \cup B) \cup A'$ (d) $(A \cup A') \cap B$
   (e) $A \cup (B \cap B')$ (f) $A \cup (B \cup B')$ (g) $A \cap (B \cap B')$
   [(a) A (b) C (c) V (d) B (e) A (f) V (g) $\emptyset$]

4. Draw Venn diagrams to demonstrate that
   (a) $(A \cup B)' = A' \cap B'$ (b) $(A \cap B)' = A' \cup B'$
   (c) $(A \cap B) \cup (A \cap B') = A$ (d) $[(A \cap B) \cap (A' \cap B')]' = V$
   (e) $A \cup (A' \cap B) = A \cup B$ (f) $A \cap (B \cup C) = (A \cap B) \cup (A \cap C)$
   (g) $(A \cap B) \cup (A \cap B') \cup (A' \cap B) = A \cup B$
5. Draw Venn diagrams to test which of the following sets are equal.
   (a) $A \cap B'$ (b) $(A \cap B)'$ (c) $(B \cup A')'$ (d) $A' \cup B'$.
   [Sets (a) and (c) and sets (b) and (d) are equal.]

4.4 Subsets of the Set of Real Numbers

The system of real numbers forms a set which is closed under the operations of addition, subtraction, multiplication, and division. The elements of the set cannot be listed. The postulates for the real numbers, as stated in Section 1.2, are criteria to determine which numbers are elements of the set.

As illustrated by the family tree of the real numbers (Table 1.1), the real numbers may be divided into a number of classes, the natural numbers, the integers, the rationals and the irrationals which are proper *subsets* of the set of real numbers.

The subsets of the set of real numbers comprise the following sets:

$N = \{\text{natural numbers}\} = \{1, 2, 3, 4, \ldots\}$

$Z = \{\text{integers}\} = \{\ldots, -3, -2, -1, 0, 1, 2, 3, \ldots\}$

$R = \{\text{rational numbers}\} = \{x = p/q : p \text{ and } q \text{ are integers and } q \neq 0\}$

$I = \{\text{irrational numbers}\} = \{x : x \text{ is a non-terminating, non-recurring decimal}\}$

Write $V = \{\text{real numbers}\}$, so that V is the universal set of which N, Z, R and I are proper subsets.

The set of natural numbers N is also a subset of the set of integers Z, i.e.

$N \subset Z$ or $\{1, 2, 3, 4, \ldots\} \subset \{\ldots, -3, -2, -1, 0, 1, 2, 3, \ldots\}$

The set of integers Z is a subset of the set of rationals R, i.e.

$Z \subset R$

The set of rationals R and the set of irrationals I are both subsets of the set of real numbers V, i.e.

$R \subset V$ and $I \subset V$

To summarize,

$N \subset Z \subset R \subset V$ and $I \subset V$

It has been pointed out in Section 1.6 that certain of these subsets obey some but not all of the postulates of the real number system as stated in Section 1.2 and that the subsets differ in their closure properties.

The set of natural numbers N is closed under addition and multiplication but is not closed under subtraction and division.

The set of integers Z is closed under addition, subtraction and multiplication but not under division.

The set of rationals R and the set of real numbers V are closed under all four operations.

EXAMPLES 4.2

(A) What is the Lowest Common Multiple (LCM) of 5 and 7?

{multiples of 5} = {5, 10, 15, 20, 25, 30, $\underline{35}$, 40, . . . 65, $\underline{70}$, 75, . . .}

{multiples of 7} = {7, 14, 21, 28, $\underline{35}$, 42, . . . 63, $\underline{70}$, 77, . . .}

Multiples of both 5 and 7 = {$\underline{35}$, 70, 105, . . .}

LCM of 5 and 7 = 35

(B) List {divisors of 36} and {divisors of 63}. Hence find the Highest Common Factor (HCF) of 36 and 63.

{divisors of 36} = {1, 2, 3, 4, 6, $\underline{9}$, 12, 18, 36}

{divisors of 63} = {1, 3, 7, $\underline{9}$, 21, 63}

HCF of 36 and 63 = 9

EXERCISES 4.3

Exercises 1–6 below relate to the following pairs of sets:

(i) $N = \{1, 2, 3, 4, \ldots\}$ $E = \{2, 4, 6, 8, \ldots\}$

(ii) $T = \{\text{positive integral multiples of 3}\} = \{3, 6, 9, 12, \ldots\}$
       $F = \{\text{positive integral multiples of 5}\} = \{5, 10, 15, 20, \ldots\}$

(iii) $A = \{\text{positive integers} \leqslant 6\} = \{1, 2, 3, 4, 5, 6\}$
       $B = \{\text{positive integers} \geqslant 4\} = \{4, 5, 6, 7, \ldots\}$

(iv) $P = \{\text{prime numbers}\}$ $O = \{\text{odd numbers}\}$

(v) $C = \{a, b, c, d, e, f, g\}$ $D = \{a, d, e, f, h\}$

(vi) $G = \{2, 2^2, 2^3, 2^4\}$ $H = \{4, 4^2, 4^3, 4^4\}$

1. Illustrate each of the above pairs of sets by Venn diagrams stating appropriate universal sets in each case.

2. List the members of the intersection and the union of each of the above pairs of sets.

3. List the members of (i) $A \cap B'$, (ii) $P \cap O'$ taking N as the universal set.
   [(i) {1, 2, 3}, (ii) {2}]

4. Express the following sets in terms of A, E, and T.
   (i) $\{3, 6\}$ (ii) $\{2, 4, 6\}$ (iii) {positive integers which are multiples of 6}
   [(i) $A \cap T$ (ii) $A \cap E$ (iii) $E \cap T$]

5. List the elements of $T \cap F$. What is the connection between $T \cap F$ and the LCM of 3 and 5?

6. Which pairs of the above 12 sets may be connected by the symbol $\subset$?

7. Let $Q = \{$divisors of 30$\}$ and $S = \{$divisors of 42$\}$. List the members of Q, R and $Q \cap R$. What is the HCF of 30 and 42?

8. Describe the following sets:
   (i) {integers which are less than 5 and also greater than 10}
   (ii) {positive integers whose cubes are greater than 7 and less than 126}

9. Is the set of negative rational numbers closed under (*a*) addition, (*b*) subtraction, (*c*) multiplication?

10. Do the following equations have solutions in the set of real numbers and, if so, in which subset?
   (*a*) $2x - 4 = 0$ (*b*) $2x + 3 = 0$ (*c*) $x^2 - 3 = 0$ (*d*) $2x^2 + 3 = 0$

4.5 The Number of Elements in a Set

In several applications of sets, particularly in probability theory, it is necessary to know the number of elements in the sets. Let us denote the number of elements in the set A by $n(A)$. Similarly, the numbers of elements which are in the union and the intersection of sets A and B will be denoted by $n(A \cup B)$ and $n(A \cap B)$ respectively.

Let

$$n(A) = n_1 + n_2 \quad \text{and} \quad n(B) = n_2 + n_3$$

Let

$$n(A \cap B) = n_2$$

These sets are illustrated in the Venn diagram (Fig. 4.7).

$$n(A \cup B) = n_1 + n_2 + n_3 = n(A) + n(B) - n(A \cap B)$$

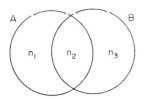

FIGURE 4.7

Venn diagrams may be conveniently used to solve problems which involve the union and intersection of a number of sets.

EXAMPLE 4.3

At an athletics meeting, identification discs, coloured red, white and blue, are distributed amongst 100 athletes. 45 athletes each receive a red disc, 45 receive a white disc, and 60 receive a blue disc. 15 athletes receive 1 red and 1 white disc, 25 receive 1 white and 1 blue, and 20 receive 1 red and 1 blue, whilst 5 athletes receive 1 disc of each colour.

(i) How many receive no disc? (ii) How many receive 1 disc only? (iii) How many receive 2 discs only? (iv) How many receive 1 white disc but no blue disc?

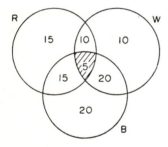

FIGURE 4.8

Figure 4.8 is a Venn diagram for the 3 sets.

R = {athletes who receive red discs} $n(R) = 45$
W = {athletes who receive white discs} $n(W) = 45$
B = {athletes who receive blue discs} $n(B) = 60$

The intersection of the sets R and W will comprise those athletes who have 1 red and 1 white disc. Therefore

$n(R \cap W) = 15$

Similarly

$n(W \cap B) = 25$
$n(R \cap B) = 20$

The intersection of R, W and B will comprise those athletes who have 1 red, 1 white and 1 blue disc. Therefore

$n(R \cap W \cap B) = 5$

This is represented by the shaded area in Fig. 4.8. By working backwards through the list of data it is now easy to determine the numbers of elements in each region of the diagram, as indicated in Fig. 4.8 from which the questions posed can be answered, thus:

(i) The total number of athletes who receive 1 or more discs
$= 45 + 20 + 20 + 10 = 95$
Since there are 100 athletes, 5 receive no disc.
(ii) The number of athletes who receive 1 disc only
$= 15 + 20 + 10 = 45$
(iii) The number of athletes who receive exactly 2 discs
$= 15 + 10 + 20 = 45$
(iv) The number of athletes who receive 1 white disc but no blue
$= 10 + 10 = 20$

EXERCISES 4.4

1. In a class of 30 pupils, each pupil plays at least one of the games cricket and football. 17 play cricket and 19 play football. How many play both cricket and football?
[6]

2. In an audience of 260 people, there are 100 men. One eighth of the women smoke and the number of men who do not smoke is one half of the number of the people who do smoke. How many of the men smoke?
[60]

3. In a group of 50 students, the numbers studying languages were found to be French, 14; German, 15; Spanish, 21; French and German, 4; French and Spanish, 5; German and Spanish, 3; all three languages, 2.
   (a) How many students were studying no language?
   (b) How many were studying Spanish only?
   (c) How many were studying Spanish and German only?
   [(a) 10 (b) 15 (c) 1]

4. Each of 135 schoolboys collects at least one of the following: stamps, coins, or autographs. 70 collect stamps, 80 collect autographs; 35 collect stamps and coins, 30 collect coins and autographs, and 25 collect autographs and stamps; 35 collect autographs only. How many collect (a) all three, (b) coins, (c) coins only?
   [(a) 10 (b) 65 (c) 10]

5. Each of a group of university students takes at least one of the subjects mathematics, physics and chemistry. 60 take physics, 30 take mathematics, and 50 take chemistry; 10 take mathematics and physics, 25 take physics and chemistry, and 20 take mathematics and chemistry. How many students take all three subjects and what is the number of students in the group?
   [5, 90]

6. 50 of a group of 120 people are women. All the men are over 21 and there are 90 people over 21. There are 20 married women of whom 5 are over 21. There are 15 married people over 21.

    (*a*) How many of the people are married?

    (*b*) How many unmarried women are over 21?

    (*c*) How many unmarried men are under 21?

    (*d*) How many men are married?

    (*e*) How many people are under 21?

[(*a*) 30   (*b*) 15   (*c*) 0   (*d*) 10   (*e*) 30]

5 The number line

5.1 The Real Line

Let us choose an arbitrary origin O on a line and measure off equal intervals of a convenient length on this line from O to the left and right of O. These intervals will be called *unit intervals* (Fig. 5.1).

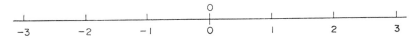

FIGURE 5.1 *The real line.*

With reference to the origin O and the chosen unit interval, a one-to-one correspondence may be established between the real numbers and the points of the line, i.e. every real number may be represented by one and only one point of the line and every point of the line will correspond to one and only one real number.

Let the origin O represent the number zero. We now adopt the convention that all points to the *right* of O represent *positive* numbers and all points to the *left* of O represent *negative* numbers, i.e. the sign of a number is associated with a definite direction from O.

In terms of the unit interval, which is taken to represent the magnitude 1, a specified number may be represented by a definite length on the line. If the distance is laid off from O in the direction associated with the sign of the given number, a *unique* point of the line, representing the number, will be determined.

The end points of the unit intervals to the right of O represent successively, as we move to the right from O, the positive integers, 1, 2, 3, ... Similarly,

FIGURE 5.2

those to the left of O represent successively, as we move to the left from O, the negative integers, $-1, -2, -3, \ldots$.

Any positive number x will be represented by a point x units to the right of O and its negative $-x$ by a point x units to the left of O.

The representative point for any *rational* number may be found *exactly*. For example, the rational number $-2\frac{5}{6}$ will be represented by a point which lies in the third unit interval to the left of O. We now divide the third unit interval into six equal subdivisions. The *exact* position of the point which represents $-2\frac{5}{6}$ is at precisely the fifth subdivision to the left of -2 (Fig. 5.2). The representative point of any rational number may be precisely determined by a limited number of subdivisions.

By definition, an irrational number cannot be expressed in the form p/q (p and q being integers). Consequently, its magnitude cannot be represented as a rational multiple of the unit interval. It follows that the *exact* position of its representative point cannot be determined. It is for this reason that the position of such points is said to be incommensurable or "unmeasurable".

An irrational number can, however, be expressed as a non-terminating and non-recurring decimal. The *approximate* position of the representative point of a given irrational number may be determined by using its decimal equivalent. For example, in the case of the irrational number $\sqrt{2} = 1.41421\ldots$, we place its representative point in the interval between the points representing 1.4 and 1.5. After subdividing this interval into tenths, we place the point representing $\sqrt{2}$ between those representing 1.41 and 1.42. We continue this process by subdividing the last interval and we place the

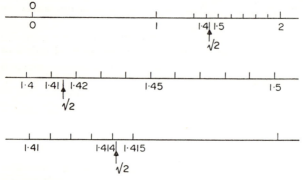

FIGURE 5.3

point representing $\sqrt{2}$ in the interval between the points representing 1.414 and 1.415, and so on (Fig. 5.3).

By continuing the process of subdivision into tenths, the representative point for the irrational number $\sqrt{2}$ may be located in as small an interval as we please, i.e. to any desired degree of accuracy.

We may, without ambiguity, refer to "the point $-2\frac{5}{6}$" as an abbreviation for "the point representing the real number $-2\frac{5}{6}$". An essential distinction between the rational and the irrational numbers has been emphasized, namely, that whilst every rational and irrational number may be represented by a unique point on the line, only the rationals may be placed *exactly* on the line.

In this Section, the methods by which the various subsets of the real numbers may be put in one-to-one correspondence with the points on a line have been described.

The line is referred to as the *real number line* or simply the *real line*.

5.2 Order and Denseness

In Chapter 1, reference was made to a special property of the set of real numbers which has not yet been discussed—the property of *order*. This property has been implicit in the principles used to represent numbers on the real line.

The natural numbers may clearly be ordered since for any distinct pair of natural numbers selected, one is less than the other, e.g.

$$2 < 4 \qquad 21 < 49$$

Rational numbers may also be ordered, since for any two distinct common fractions selected, one is less than the other, e.g.

$$\frac{1}{4} < \frac{3}{8} \qquad \frac{11}{16} < \frac{3}{4}$$

If the natural numbers are arranged in order of magnitude, they form an ordered array

$$1, 2, 3, 4, 5, 6, 7, \ldots$$

in which each member has a *definite* next larger and next smaller member.

This cannot be done with fractions since, between any two distinct fractions, another fraction may always be inserted, the magnitude of which is intermediate between the magnitudes of the first two. Thus we may insert

$\dfrac{21}{32}$ between $\dfrac{5}{8}$ and $\dfrac{11}{16}$, $\left(\dfrac{5}{8} < \dfrac{21}{32} < \dfrac{11}{16}\right)$, and

$\dfrac{43}{64}$ between $\dfrac{21}{32}$ and $\dfrac{11}{16}$, $\left(\dfrac{21}{32} < \dfrac{43}{64} < \dfrac{11}{16}\right)$

This process may be continued indefinitely. In these examples, the fraction inserted has the average magnitude of the two given fractions. In general terms, we insert the fraction

$$\frac{1}{2}\left(\frac{p}{q} + \frac{r}{s}\right) = \frac{ps + qr}{2qs}$$

between the fractions p/q and r/s. This property of the fractions is described by saying that the set of fractions is dense, implying that between any two rational numbers another rational number may be found.

Whilst the set of rational numbers is dense, it cannot be said to be "complete" in the sense that there are certain points on the real line which cannot be labelled by any rational number. These are the points which represent the irrational numbers. In other words, there are more points on the real line than there are rational numbers.

The concept of the *complete ordering* of the real numbers, which will not be discussed here, means that every point on the real line represents a unique number, rational or irrational, and vice versa. That is, there are no "gaps" on the real line.

DEFINITION 5.1
If x and y are real numbers, we define $x > y$ if and only if $x - y$ is a positive number and $x > y$ may also be written $y < x$.

This means that, on the real line, the point x lies to the *right* of the point y if $x > y$ and the point x lies to the *left* of the point y if $x < y$.

The property of order possessed by the set of real numbers is expressed by the addition of the following postulate to the group stated in Section 1.2.

POSTULATE 8:   *The Property of Order*
For any two real numbers x and y, one and only one of the following relations is true

    (i) $x > y$ (ii) $x = y$ (iii) $x < y$.

5.3   *The Representation of a Set of Numbers on the Real Line*

The set of all real numbers which have values between 2 and 4, i.e. $\{x : 2 < x < 4\}$, may be represented by a segment or *interval* of the number line

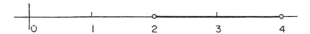

FIGURE 5.4

between the representative points for $x = 2$ and $x = 4$ as in Fig. 5.4. The open circles at the ends of the interval indicate that the numbers 2 and 4 are not included in the set.

The graph of the set $\{x : 2 \leqslant x \leqslant 4\}$ is the same line segment but the small circles at the ends of the interval are now filled in to indicate the inclusion of the numbers 2 and 4, as in Fig. 5.5.

FIGURE 5.5

If a and b are any real numbers, the intervals which represent the sets

$$\{x : a < x < b\} \quad \text{and} \quad \{x : a \leqslant x \leqslant b\}$$

are called *open* and *closed* intervals respectively and are denoted by (a, b) and $[a, b]$ respectively. Intervals may be open (or closed) at both ends or at one end only, and $(a, b]$ denotes an interval which is open on the left and closed on the right.

The intersection and union of sets of numbers may be represented on the number line as illustrated in Fig. 5.6.

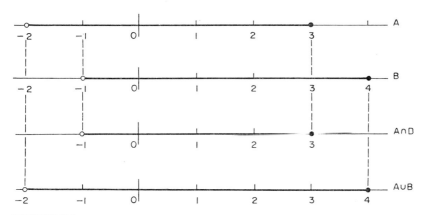

FIGURE 5.6
$A = \{x : -2 < x \leqslant 3\}$ $B = \{x : -1 < x \leqslant 4\}$
$A \cap B = \{x : -1 < x \leqslant 3\}$ $A \cup B = \{x : -2 < x \leqslant 4\}$

5.4 *Mathematical Sentences*

In the English language, the sentence "Charles I was an English king" becomes an open sentence if we replace "Charles I" by a blank. This open sentence becomes true or false according to the name of the king which replaces the blank. If the replacement is "Louis XIV", the statement is false but if the replacement is "Henry VIII", it is true. Here the set of replacements or universal set for this open sentence may be the set of all past English kings.

The "language of mathematics" enables us to express ideas by using words and symbols comprising "mathematical sentences". A mathematical sentence must be such that one can definitely decide that it is either true or false.

The following are examples of mathematical sentences:

(i) $9 < 11$ (true)

(ii) Open sentences:

    (*a*) $x + 2 = 6$ (*b*) $x \geqslant 4$

    (*c*) $x^2 = 9$ (*d*) $x^2 = -9$

In (ii), x is a "placeholder" for some number or is a pronumeral (cf. pronoun) drawn from a specified universal set V. For example, if V is the set of integers, the only replacement for x which makes (*a*) true is $x = 4$. In (*b*) the replacement set which makes the open sentence $x \geqslant 4$ true is the set of integers $\{4, 5, 6, 7, \ldots\}$ or $\{x : x \geqslant 4$ and x is an integer$\}$.

In (*c*) the only replacement values are $x = +3$ or -3. In (*d*) there is no possible value of x in the set of integers which makes $x^2 = -9$ true, so that the replacement set is $\varnothing$, the empty set.

These replacement sets are called the *solution sets* or *truth sets* of the open sentences. The open sentence thus plays the role of a selector from V and divides V into two sets, one for which the sentence is true and the other for which it is false.

5.5 *Rules for Inequalities*

Using Definition 5.1 and Postulate 8, we prove some important results for inequalities.

Let a, b, c be any real numbers.

We shall assume it to be axiomatic that if $a > 0$ and $b > 0$, then $a + b > 0$ and $ab > 0$.

(i) To PROVE that if $a > b$ and $b > c$, then $a > c$.

$$a > b \Rightarrow a - b > 0$$
$$b > c \Rightarrow b - c > 0$$

Therefore

$$(a - b) + (b - c) > 0$$
$$a - c > 0$$
$$a > c$$

(ii) To PROVE that if $a > b$ then $a + c > b + c$.

$$(a + c) - (b + c) = a - b$$

But $a - b > 0$, therefore

$$(a + c) - (b + c) > 0$$
$$a + c > b + c$$

(iii) To PROVE that if $a > b$ and $c > 0$ then $ac > bc$ but if $c < 0$ then $ac < bc$.

$$ac - bc = c(a - b) \quad \text{and} \quad a - b > 0$$

Therefore

$$\text{if } c > 0, \ c(a - b) > 0 \quad \text{then } ac - bc > 0 \qquad ac > bc$$
$$\text{if } c < 0, \ c(a - b) < 0 \quad \text{then } ac - bc < 0 \qquad ac < bc$$

The converses of (ii) and (iii) may be similarly proved.

Summarising, the rules for inequality and equality are:

Rule 1: $a \geqslant b \Leftrightarrow a + c \geqslant b + c$
Rule 2: $a \geqslant b \Leftrightarrow ac \geqslant bc \quad \text{if} \quad c > 0$
Rule 3: $a \geqslant b \Leftrightarrow ac \leqslant bc \quad \text{if} \quad c < 0$

The *graphs of solution sets* of open sentences in one variable may be drawn on the number line as described in Section 5.3.

EXAMPLE 5.1

Find the solution sets for the following inequalities.

(A) $\dfrac{1}{3} x - 3 \leqslant 5$

$$\frac{1}{3}x - 3 + 3 \leqslant 5 + 3 \quad \text{(Rule 1)}$$

$$\frac{1}{3}x \leqslant 8$$

$$x \leqslant 24 \quad \text{(Rule 2)}$$

Solution set $\{x : x \leqslant 24\}$ (Fig. 5.7(*a*))

(B) $3(4x - 9) > 2x - 7$

$\ 12x - 27 > 2x - 7$

$\ 12x - 27 + 27 - 2x > 2x - 7 + 27 - 2x$ (Rule 1)

$\ 10x > 20$

$\ x > 2$ (Rule 2)

Solution set $\{x : x > 2\}$ (Fig. 5.7(*b*))

(C) $2(x - 1) - 7 \geqslant 2x - 9$

$\ 2x - 9 \geqslant 2x - 9$

If the equality sign is taken, the solution set is the *universal* set whilst if the inequality sign is taken, the solution set is the *empty* set. When both signs are taken, the solution set is the *universal* set.

(*a*)

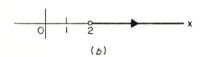

(*b*)

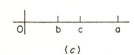

(*c*)

FIGURE 5.7

CHAINS OF INEQUALITIES

Earlier in this section, we saw that

$$a > b,\, b > c \ (\text{i.e. } a > b > c) \Rightarrow a > c$$

If $a > b$, $a > c$, it is not necessarily true that $a > b > c$. The case for which $a > c > b$ is illustrated in Fig. 5.7(*c*).

EXAMPLE 5.2

Find the solution set for the following chain of inequalities

$$6x - 3 < 2x - 5 < 4 + 7x$$

We have (i) $6x - 3 < 2x - 5$ and   (ii) $2x - 5 < 4 + 7x$

$$4x < -2 \qquad\qquad\qquad -9 < 5x$$
$$x < -0.5 \qquad\qquad\qquad -1.8 < x$$

Solution set $A = \{x : x < -0.5\}$. Solution set $B = (x : x > -1.8)$. The solution set for the chain of inequalities is $A \cap B$, which is $\{x : -1.8 < x < 0.5\}$. The graph of this set is shown in Fig. 5.8.

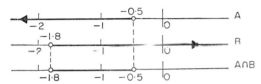

FIGURE 5.8
$A = \{x : x < -0.5\}$ $B = \{x : x > -1.8\}$ $A \cap B = \{x : -1.8 < x < -0.5\}$

QUADRATIC INEQUALITIES

If $a > b$, it does not necessarily follow that $a^2 > b^2$. For example,

$$2 > -3 \quad \text{but} \quad (2)^2 < (-3)^2$$

It may, however, be readily proved that if $a > b > 0$ then $a^2 > b^2$.

$$a > b > 0 \Rightarrow a^2 > ab \quad \text{since} \quad a > 0 \quad \text{(Rule 2)}$$
$$a > b > 0 \Rightarrow ab > b^2 \quad \text{since} \quad b > 0 \quad \text{(Rule 2)}$$

Therefore

$$a^2 > ab > b^2 \qquad \text{i.e. } a^2 > b^2$$

Similarly, it may be proved that if $b < a < 0$ then $a^2 < b^2$.

EXAMPLE 5.3

Find the solution sets for the following.

(A) (a) $2x^2 = 18$ (b) $2x^2 < 18$ (c) $2x^2 \leqslant 18$.

(a) $2x^2 = 18$ $x^2 = 9$ $x = \pm 3$
   Solution set $\{+3, -3\}$ (Fig. 5.9(a))
(b) $2x^2 < 18$ $x^2 < 9$ $x < +3$ and $x > -3$
   Solution set $\{x : -3 < x < +3\}$ (Fig. 5.9(b))
(c) $2x^2 \leqslant 18$ Solution set $\{x : -3 \leqslant x \leqslant +3\}$ (Fig. 5.9(c))

(a)
$$-3 \quad 0 \quad +3 \quad\quad x$$

(b)
$$-3 \quad 0 \quad +3 \quad\quad x$$

(c)
$$-3 \quad 0 \quad +3 \quad\quad x$$

FIGURE 5.9
(a) $2x^2 = 18$ (b) $2x^2 < 18$ (*open interval*)   (c) $2x^2 \leqslant 18$ (*closed interval*)

(B) (a) $(2x - 5)(3x + 4) = 0$
(b) $(2x - 5)(3x + 4) < 0$
(c) $(2x - 5)(3x + 4) \leqslant 0$

(a) $(2x - 5)(3x + 4) = 0 \Leftrightarrow (2x - 5) = 0$ or $(3x + 4) = 0$
i.e. $x = 2\frac{1}{2}$ or $-1\frac{1}{3}$
Solution set $\{2\frac{1}{2}, -1\frac{1}{3}\}$

(b) The points $2\frac{1}{2}$ and $-1\frac{1}{3}$ divide the number line into three regions $A = \{x : x \leqslant -1\frac{1}{3}\}$, $B = \{x : -1\frac{1}{3} < x < 2\frac{1}{2}\}$, and $C(x : x \geqslant 2\frac{1}{2})$ as shown in Fig. 5.10(a).
Let us find the sign of $(2x - 5)(3x + 4)$ in the regions A, B and C by considering the signs of the factors $(2x - 5)$ and $(3x + 4)$ in the same regions

Sign of	A	B	C
$(2x - 5)$	—	—	+ (or 0)
$(3x + 4)$	— (or 0)	+	+
$(2x - 5)(3x + 4)$	+ (or 0)	—	+ (or 0)

It appears that $(2x - 5)(3x + 4) < 0$ only if x has values in the range B. Solution set is $\{x : -1\frac{1}{3} < x < 2\frac{1}{2}\}$

(c) The solution set for $(2x - 5)(3x + 4) \leqslant 0$ is $\{x : -1\frac{1}{3} \leqslant x \leqslant 2\frac{1}{2}\}$ as illustrated in Fig. 5.10(b).

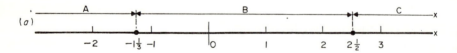

(a)
$$-2 \quad -1\frac{1}{3} \; -1 \quad 0 \quad 1 \quad 2 \quad 2\frac{1}{2} \quad 3 \quad\quad x$$

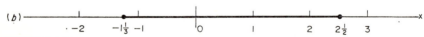

(b)
$$-2 \quad -1\frac{1}{3} \; -1 \quad 0 \quad 1 \quad 2 \quad 2\frac{1}{2} \quad 3 \quad\quad x$$

FIGURE 5.10

5.6 Absolute Value

The absolute value of a real number is a useful concept. It may appropriately be considered here since its applications are usually concerned with the property of order.

On the number line, a real number x is represented by a point placed x units from the origin: to the right of the origin if x is positive and to the left if x is negative.

The absolute value of the number x is represented by the distance of the point x from the origin without regard to sign.

DEFINITION 5.2
The absolute value of a real number x, denoted by $|x|$, is defined by

$$|x| = \begin{cases} x & \text{if } x \geqslant 0 \\ -x & \text{if } x < 0 \end{cases}$$

From this definition, it is seen that

$$|3| = |-3| = 3 \qquad |101| = |-101| = 101 \qquad |0| = 0$$

In effect, the sign of a number is ignored.

For any real number x, $|x| = |-x|$.

Since $y - x = -(x - y)$, it follows that

$$|x - y| = |y - x|$$

The distance d between any two points x and y on the number line (Fig. 5.11) may be expressed

$$d = |x - y| = |y - x|$$

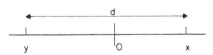

FIGURE 5.11

Some important results for absolute values follow:
  (i) $|xy| = |x|\,|y|$
  (ii) $|x/y| = |x|/|y|$ provided $y \neq 0$
  (iii) $|x + y| \leqslant |x| + |y|$
  (iv) $|x - y| \geqslant |x| - |y|$

from which it follows that

$$|x - y| \geqslant |y| - |x|$$

and hence

$$|x - y| \geqslant ||x| - |y||$$

(i) If x and y are both positive

$$|xy| = xy = |x|\,|y|$$

If x and y are both negative

$$|xy| = xy = (-x)(-y) = |x|\,|y|$$

If either x or y is positive and the other is negative, say $x > 0, y < 0$, then

$$|xy| = -xy = (x)(-y) = |x|\,|y|$$

(iii) If x and y are both positive or both negative or if x or $y = 0$, then

$$|x + y| = |x| + |y|$$

e.g.

$$|5 + 3| = 8 \quad \text{and} \quad |5| + |3| = 8$$
$$|(-4) + (-2)| = |-6| = 6 \quad \text{and} \quad |-4| + |-2| = 4 + 2 = 6$$

If x and y have opposite signs,

$$|x + y| < |x| + |y|$$

e.g. If $x = 5, y = -3$

$$|5 + (-3)| = 2 \quad \text{and} \quad |5| + |-3| = 5 + 3 = 8$$

It will be left to the reader to verify (ii) and (iv).

EXAMPLES 5.4

(A) Find the solution sets of the following inequalities.

(a) $|x| < 5$
Let $x = X$ be a solution.
    (i) Let $X > 0$ $|x| = X < 5$
    (ii) Let $X = 0$ $|x| = 0 < 5$ (inequality satisfied by $x = 0$)
    (iii) Let $X < 0$ $|x| = -X < 5$, i.e. $X > -5$

Therefore

$$-5 < X < 5$$

i.e.

$$-5 < x < 5$$

Solution set is $\{x: -5 < x < 5\}$
In general, if $|x| < K$, $K > 0$,
  Solution set is $\{x: -K < x < K\}$

(b) $|x - 3| < 6$
It follows from (a) that if $|x - 3| < 6$ then

$$-6 < (x - 3) < 6$$

i.e.

$$-3 < x < 9$$

Solution set is $\{x: -3 < x < 9\}$

(B) Find the solution set for the inequality

$$|x - 3| + |x + 3| < 8$$

Let $x = X$.
(i) When $X \geqslant 3$, $X - 3 \geqslant 0$ and $X + 3 > 0$.

$$(X - 3) + (X + 3) < 8$$
$$2X < 8$$
$$X < 4$$

Inequality is true for $3 \leqslant X < 4$
(ii) When $X \leqslant -3$, $X - 3 \leqslant 0$ and $X + 3 \leqslant 0$.

$$-(X - 3) - (X + 3) < 8$$
$$-2X < 8$$
$$X > -4$$

Inequality is true for $-4 < X \leqslant -3$
(iii) When $-3 < X < 3$, $X - 3 < 0$ and $X + 3 > 0$.

$$-(X - 3) + (X + 3) < 8$$
$$6 < 8$$

Inequality is true for $-3 < X < 3$
Solution set is $\{x: -4 < X < 4\}$

(C) Find the solution set for the inequality

$$|4x + 5| + |3x - 1| < 7$$

Let $x = X$.
(i) When $X \geqslant 1/3$, $4X + 5 > 0$ and $3X - 1 \geqslant 0$.

$$4X + 5 + 3X - 1. < 7$$
$$7X + 4 < 7$$
$$7X < 3$$
$$X < \frac{3}{7}$$

Inequality is true for $1/3 \leqslant X < 3/7$
(ii) When $X \leqslant -5/4$, $4X + 5 \leqslant 0$ and $3X - 1 < 0$.

$$-(4X + 5) - (3X - 1) < 7$$
$$-7X - 4 < 7$$
$$-7X < 11$$
$$X > -\frac{11}{7}$$

Inequality is true for $-11/7 < X \leqslant -5/4$
(iii) When $-5/4 < X < 1/3$, $4X + 5 > 0$ and $3X - 1 < 0$

$$4X + 5 - (3X - 1) < 7$$
$$X + 6 < 7$$
$$X < 1 \quad \text{which is true}$$

Solution set is $\{x : -11/7 < x < 3/7\}$

EXERCISES 5.1
1. Represent the following sets on the number line:
   (a) $\{x : x \geqslant -2\} \cap \{x : x < 3\}$
   (b) $\{x : 2 < x \leqslant 5\} \cap \{x : 3 \leqslant x < 6\}$
   (c) $\{x : -4 \leqslant x < 3\} \cup \{x : -1 < x < 4\}$
   (d) $\{x : x < -1\} \cap \{x : x > 2\}$
2. Find the solution sets of the following inequalities and represent them on the number line.
   (a) $7x > 3x + 8$
   (b) $4 - 4(x - 5) > 2(2 - x) - 6$
   (c) $\frac{1}{5}(2x - 1) - \frac{1}{2}(3x + 1) < \frac{2}{5}$
   (d) $\frac{1}{3}(2x - 1) + \frac{1}{5}(4x + 1) < \frac{3}{4}(x - 4)$
   [(a) $x > 2$ (b) $x < 13$ (c) $x > -1$ (d) $x < -4$]

3. Find the solution sets of the following chains of inequalities
   (a) $5 - 2x < 11 < x - 3$
   (b) $3x - 2 < x < 6 + 4x$
   (c) $x - 3 \leqslant 3x - 5 \leqslant x + 7$
   (d) $3x - 1 \geqslant 6 - 5x \geqslant 14 - 4x$
   [(a) $x > 14$ (b) $-2 < x < 1$ (c) $1 \leqslant x \leqslant 6$ (d) No solution set]

4. Represent the solution sets of the following inequalities on the number line
   (a) $(x + 4)(x - 3) \leqslant 0$
   (b) $(5 - x)(x - 7) \geqslant 0$
   (c) $(3x + 1)(2x - 5) < 0$
   (d) $(x - 1)(x - 2)(x - 3) < 0$
   [(a) $-4 \leqslant x \leqslant 3$ (b) $5 \leqslant x \leqslant 7$ (c) $-\frac{1}{3} < x < 2\frac{1}{2}$ (d) $x < 1$ or
   $2 < x < 3$]

5. Find the solution sets of the following inequalities
   (a) $|x - 4| = 8$
   (b) $|x + 5| \leqslant 7$
   (c) $|x| + |x - 6| < 10$
   (d) $|2x - 3| + |3x + 4| \leqslant 11$.
   [(a) $x = 12, -4$ (b) $-12 \leqslant x \leqslant 2$ (c) $-2 < x < 8$ (d) $-(12/5) \leqslant x \leqslant 2$]

6 Relations and functions

6.1 Relations within a Set

Each of the following sentences represents a binary relation between two individuals:

Tom is taller than John

Mary and Joan are sisters

In mathematics, many such binary operations are in common use, e.g.

(a) The line AB is parallel (perpendicular) to the line CD.
(b) The triangle ABC is congruent (similar) to the triangle PQR.
(c) 24 is a multiple of (is divisible by) 8.
(d) x is greater (less) than y (x, y integers).

In a large number of relations in mathematics, the elements which are connected by the relation belong to the *same* set. Examples (a) and (b) may be taken to represent relations between the elements of a set of lines and between the elements of a set of triangles in a plane respectively. In (c) and (d), we are concerned with relations between the elements of the set of integers.

In such instances, we have a relation *within* a set. Let A be a set of triangles in a plane. Any pair of triangles x and y will be *either* similar *or* not similar. The relation of similarity may be said to divide the set A into two subsets so that the relation is true for pairs of elements of one subset but is false for pairs of elements of the second subset.

Let R be the relation which denotes that triangles x and y are similar, i.e. that x and y are in the same subset. R may be defined by

$$R = \{(x, y): x, y \in A \text{ and } x \text{ is similar to } y\}$$

We formally define a *relation* as follows:

DEFINITION 6.1
A set A is said to have a relation R if for each pair (x, y) of elements of A, the phrase "x is in relation R to y" has a meaning and is true or false depending only upon the selection of x and y.

We use the notation xRy to signify that "x is in relation R to y" and $x\bar{R}y$ to signify the negation of the phrase. For any pair of elements x, y *either xRy or $x\bar{R}y$* is true but *not both*.

6.2 Equivalence Relations

Many relations on a set are characterised by three important properties which are defined as follows:

DEFINITION 6.2
Let R be a relation on a set A, then
(1) R is a *reflexive* relation $\Leftrightarrow xRx$ for each $x \in A$. This states the obvious fact that every element x in A stands in any given relationship to itself, e.g. every triangle is congruent or similar to itself.
(2) R is a *symmetric* relation $\Leftrightarrow xRy \Rightarrow yRx$ for $x, y \in A$. A relation is symmetric if whenever x is in the given relationship to y, then y is in the same relationship to x, i.e. both x and y are members of the subset for which R is true.
(3) R is a *transitive* relation $\Leftrightarrow xRy$ and $yRz \Rightarrow xRz$ for $x, y, z \in A$. A relation is transitive if whenever x is in the given relationship to y and y is in the same relationship to z, then x is also in that same relationship to z, i.e. x, y and z belong to the same subset.
A relation R on a set A which is reflexive, symmetric and transitive is said to be an *equivalence relation* on A. Equality ($=$) is the most obvious example of an equivalence relation.

Consider some of the examples in Section 6.1. Let the set of lines in a plane have the relation "is parallel to". This is clearly an equivalence relation since
1. Any line x is parallel to itself (Reflexive).
2. If line $x \parallel$ line y then line $y \parallel$ line x (Symmetric).
3. If line $x \parallel$ line y and line $y \parallel$ line z then line $x \parallel$ line z (Transitive).
On the other hand, the relation "is perpendicular to" is *not* an equivalence relation since a line cannot be perpendicular to itself. This relation is, however, symmetric though it is not transitive.

Consider the relation "less than" ($<$) on the set of real numbers. This is clearly not a reflexive relation since, for example, $2 \not< 2$. It is not a symmetric relation since $2 < 3$ does not imply $3 < 2$ but it is a transitive relation since $2 < 3$ and $3 < 4 \Rightarrow 2 < 4$. The relation $\leqslant$ on the set of real numbers is a reflexive and transitive but a nonsymmetric relation. It is seen that there are relations for which one, two or three of the properties of Definition 6.2 hold.

Consider the equivalence relation "is parallel to" on a set of all lines L in a plane. This equivalence relation will partition the set L into a number of subsets $L_1, L_2, L_3, \ldots$, each subset comprising sets of lines which are parallel (Fig. 6.1). Every line in the set L will be selected as a member of one and only one of the subsets.

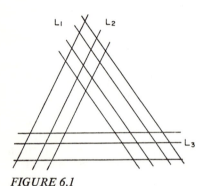

FIGURE 6.1

In general, an equivalence relation R on a set A will divide the set A into mutually exclusive subsets, i.e. every element of A will be a member of one and only one subset of A and any pair of elements x, y of the set A, such that xRy, belong to the same subset.

These subsets are called the *equivalence classes* by the equivalence relation R on the set A.

EXERCISES 6.1

In 1–8 below, which properties of Definition 6.2 do the following relations satisfy?

Set	*Relation*
1. Men in the United Kingdom	"is the father of"
2. People in the United Kingdom	(i) "has the same surname as" (ii) "live in the same town as"
3. Men	$xRy \Leftrightarrow x$ and y have at least one common parent.
4. Triangles in a plane	"is congruent to"
5. Triangles in a plane	"is similar to"
6. Integers	$xRy \Leftrightarrow (x - y)$ is divisible by 2
7. Subsets of a given set	$\subset$
8. Subsets of a given set	$ARB \Leftrightarrow A \cap B = \phi$

9. Give an example of a relation which is
   (i) reflexive and symmetric but not transitive,
   (ii) symmetric and transitive but not reflexive,
   (iii) transitive and reflexive but not symmetric.

10. Show that the following are equivalence relations and state the equivalence classes.
    (i) $aRb \Leftrightarrow (a - b)$ is divisible by 3 where $a, b \in Z$.
    (ii) $(a, b)R(c, d) \Leftrightarrow a - c = b - d$ | where (a, b), (c, d) are points in a
    (iii) $(a, b)R(c, d) \Leftrightarrow a - c = d - b$ | plane and $a, b, c, d \in Z$.
    (iv) $(a, b)R(c, d) \Leftrightarrow ad = bc$ where $a, b, c, d \in Z$ and $b \neq 0, d \neq 0$.
        Z is the set of integers.

6.3   The Cartesian Plane

We have already seen how a set of numbers may be represented by points on the number line. When we have two distinct sets of numbers, which are to be paired, we represent all possible pairs by points in a plane.

Two straight lines, Ox, Oy, are drawn at right angles through an origin O. They are called rectangular *coordinate* axes. A network of unit squares is constructed by drawing lines parallel to Ox, Oy at unit intervals apart.

A number pair, e.g. (2, 3), is represented by the point A which is the intersection of the line parallel to Oy which passes through point 2 on Ox and of the line parallel to Ox which passes through the point 3 on Oy (Fig. 6.2).

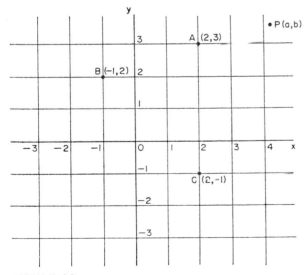

FIGURE 6.2

A is called the point (2, 3). Similarly, the points B and C represent the number pairs $(-1, 2)$ and $(2, -1)$ respectively.

In general, the number pair (a, b) is represented by the point P(a, b) which is displaced from O by a units in the direction of Ox and b units in the direction of Oy.

It will be observed that the number pairs $(-1, 2)$ and $(2, -1)$ are represented by different points B and C respectively and are therefore different. Thus the order in which the numbers of a pair are written is important.

We refer to (a, b) as an *ordered pair*. The general ordered pair represented by a point in the plane is usually written (x, y), and x and y are called the *abscissa* or x-coordinate and the *ordinate* or y-coordinate respectively, and are jointly called the *coordinates* of the representative point. Coordinates may be positive and negative. The point whose coordinates are x and y is often referred to briefly as "the point (x, y)".

The ordered pair (x, y) has a one-to-one correspondence with *one* point in the plane and vice versa.

A coordinate system of this kind is called a rectangular *Cartesian* coordinate system and was invented by the French mathematician Descartes.

6.4   Set of Ordered Pairs—Cartesian Product

We have seen how ordered pairs may be plotted on the Cartesian plane. Let us now see how such ordered pairs may arise.

We introduce a simple general concept, the *Cartesian product*. We begin with two small finite sets

$$A = \{1, 2, 3, 4\} \qquad B = \{1, 3\}$$

A set of ordered pairs may be formed by choosing the first number from A and the second number from B.

This set has 4×2, i.e. 8 elements as follows:

$$\{(1, 1), (1, 3), (2, 1), (2, 3), (3, 1), (3, 3), (4, 1), (4, 3)\}$$

This set of ordered pairs is called the Cartesian product of A and B.

DEFINITION 6.3
The *Cartesian product* of two sets A and B is the set of all ordered pairs which are formed by choosing the first member of the pair from A and the second member from B.

It is denoted by $A \times B$ (read: "A cross B").

If $B = A$, $A \times A$ is the Cartesian product of a set A with itself and is called the Cartesian product of A. For example, if $A = \{1, 2, 3\}$, the Cartesian product of A with itself is

$$A \times A = \{(1, 1), (1, 2), (1, 3), (2, 1), (2, 2), (2, 3), (3, 1), (3, 2), (3, 3)\}$$

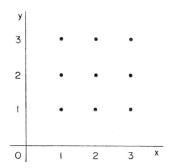

FIGURE 6.3

The graph of the set $A \times A$ is shown in Fig. 6.3 as a lattice of points. If $A = \{1, 2, 3, 4, \ldots\}$, then the set of ordered pairs $A \times A$ is represented by a lattice of all the points in the first quadrant which have integral coordinates. If V is the set of real numbers, the graph of the set $V \times V$ is the whole Cartesian plane and there is a one-to-one correspondence between every point of the plane and every element of the set $V \times V$.

6.5 Relations Between Two Sets

If we have two sets A and B and a criterion is stated by which a given element of the set A is associated with one or more elements of the set B, then we have a *relation* from the set A into the set B. If every element of the set A is associated with a *unique* element of the set B, then we have a *mapping* from the set A into the set B. The sets A and B may be the same or different. If $x \in A$ and y is an element of B which is associated with x by the relation R, we write xRy. (x, y) is an ordered pair and the relation R selects a set of ordered pairs, which is a subset of the Cartesian product $A \times B$. A relation R may thus be thought of as a set of ordered pairs comprising a subset of $A \times B$. All relations in $A \times B$ may be identified with the corresponding subsets of $A \times B$.

6.6   Sentences in Two Variables

The concept of "mathematical sentence" was introduced in Section 5.4. Let us now consider how sentences in two variables may establish a relation between two sets.

The following are familiar examples of sentences in two variables:

$$y = x + 1 \qquad y < x + 1 \qquad x^2 + y^2 \leqslant 9$$

Consider the sentence $y = x + 1$. This sentence is true for some ordered pairs and false for others. Thus, it is true for $x = 2$, $y = 3$, i.e. for the ordered pair $(2, 3)$ which is called a *solution* of this sentence. It is false for $x = 3$, $y = 2$, i.e. for the ordered pair $(3, 2)$.

We shall denote the solution set of the sentence $y = x + 1$ by

$$\{(x, y) : y = x + 1\}$$

which is read as "the set of all ordered pairs (x, y) such that $y = x + 1$".

Of course, those ordered pairs which will actually be elements of this solution set will depend upon the universal sets from which x and y may be selected. For example, if the ordered pairs are to be selected from the Cartesian set $A \times A$ where $A = \{1, 2, 3, 4\}$, the totality of ordered pairs under consideration will be represented by the lattice of 16 points shown in Fig. 6.4(a).

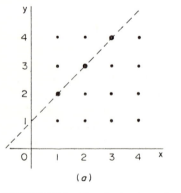
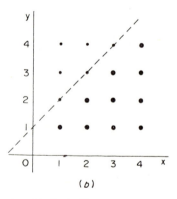

(a)	(b)

FIGURE 6.4   *Graphs in $A \times A$, where $A = \{1, 2, 3, 4\}$.*
(a) *Graph of $R_1 = \{(x, y) : y = x + 1\}$*   (b) *Graph of $R_2 = \{(x, y) : y < x + 1\}$*
    or of $y = x + 1$.   *or of $y = x + 1$.*

In $A \times A$, the solution set of the sentence $y = x + 1$ may be listed as follows:

$$\{(1, 2), (2, 3), (3, 4)\}$$

The sentence $y = x + 1$ may be regarded as the selector of a subset of ordered pairs from the Cartesian set $A \times A$ by associating elements x from the set A with elements y from an identical set A such that each element x and the corresponding element y satisfy the equation $y = x + 1$.

If R_1 is the relation which has been established by the sentence $y = x + 1$, we may identify R_1 with the subset of $A \times A$ for which $y = x + 1$ is true in $A \times A$. Thus

$$R_1 = \{(1, 2), (2, 3), (3, 4)\} - \{(x, y): y = x + 1\}$$

The graph of this solution consists of the three points (heavily marked) in Fig. 6.4(*a*). This will be called the graph of the relation $R_1 = \{(x, y): y = x + 1\}$ or simply the graph of $y = x + 1$.

For the sentence $y < x + 1$, the solution set in $A \times A$ may be listed as follows:

$$\{(1, 1), (2, 1), (2, 2), (3, 1), (3, 2), (3, 3), (4, 1), (4, 2), (4, 3), (4, 4)\}$$

and its graph consists of the ten points (heavily marked) in Fig. 6.4(*b*).

This is the graph of the relation of $R_2 = \{(x, y): y < x + 1\}$ or of $y < x + 1$.

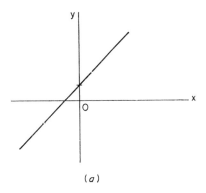

(*a*)

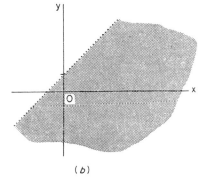

(*b*)

FIGURE 6.5 *Graphs in $V \times V$.*
(*a*) *Graph of* $R_3 = \{(x, y): y = x + 1\}$ (*b*) *Graph of* $R_4 = \{(x, y): y < x + 1\}$
  or of $y = x + 1$.   *or of* $y < x + 1$.

If V is the set of real numbers, $V \times V$ is represented by the whole plane. In $V \times V$, the solution sets of $y = x + 1$ and $y < x + 1$ are infinite sets of ordered pairs. The graph of $y = x + 1$ is a straight line R_3 (Fig. 6.5(a)) and that of $y < x + 1$ is the "half plane" R_4 below the straight line graph of $y = x + 1$ (Fig. 6.5(b)), this straight line being itself excluded.

DEFINITION 6.4

A locus is defined to be the set of those points and only those points which satisfy a given condition. This implies that a locus is the graph of a relation. If we consider only loci which lie in a plane, the condition is usually, but not necessarily, an equation or inequality in two variables.

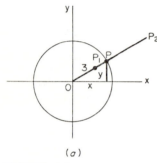

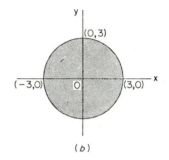

(*a*)   (*b*)

FIGURE 6.6
(a) *Graph of* $x^2 + y^2 = 9$ (b) *Graph of* $R_5 = \{(x, y): x^2 + y^2 \leqslant 9\}$.

The graph of the solution set of the equation or inequality is called the locus of the equation or inequality. Consider the condition $x^2 + y^2 \leqslant 9$. If a point $P(x, y)$ lies on the circumference of a circle centre O and radius 3 (Fig. 6.6(a)) then it is obvious that

$$x^2 + y^2 = 3^2 = 9$$

If P lies *inside* this circle then $x^2 + y^2 < 9$ and if *outside* then $x^2 + y^2 > 9$.
The condition $x^2 + y^2 \leqslant 9$ will thus select from $V \times V$ the set of ordered pairs

$$\{(x, y): x^2 + y^2 \leqslant 9\}$$

The graph of this set of ordered pairs is the circumference and interior of the circle centre O and radius 3. The set of points in this region is the locus of

$x^2 + y^2 \leqslant 9$ and is also the graph of the relation

$$R_5 = \{(x, y):x^2 + y^2 \leqslant 9\}$$

as shown in Fig. 6.6(b).
  The graphs of the relations

$$R_6 = \{(x, y):x^2 = y\} \quad \text{and} \quad R_7 = \{(x, y):y^2 = x\}$$

are shown in Fig. 6.7.

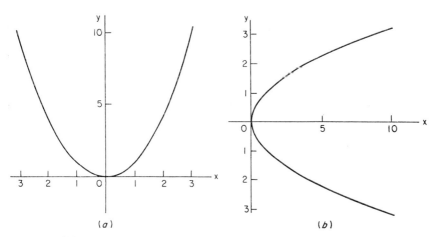

(a) (b)

FIGURE 6.7
(a) *Graph of* $R_6 = \{(x, y):x^2 = y\}$ (b) *Graph of* $R_7 = \{(x, y):y^2 = x\}$.

EXAMPLE 6.1
Draw the graphs of the solution sets, in two dimensions, of the following open sentences in $V \times V$ where V is the set of real numbers
(i) $y = 3x$ (ii) $2x + 3y = 6$ (iii) $x^2 \leqslant 4$ (iv) $y = x^2 - 1$
  Some of the ordered pairs in the solution sets of the above open sentences, corresponding to integral values of x, are tabulated below.

(i) $y = 3x$

x	-2	-1	0	1	2	3	4	5
$y = 3x$	-6	-3	0	3	6	9	12	15

(ii) $2x + 3y = 6$
  or $y = 2 - \frac{2}{3}x$

x	-2	-1	0	1	2	3	4	5
$y = 2 - \frac{2}{3}x$	$3\frac{1}{3}$	$2\frac{2}{3}$	2	$1\frac{1}{3}$	$\frac{2}{3}$	0	$-\frac{2}{3}$	$-1\frac{2}{3}$

(iv) $y = x^2 - 1$

x	-2	-1	0	1	2	3
$y = x^2 - 1$	3	0	-1	0	3	8

4

relations and functions

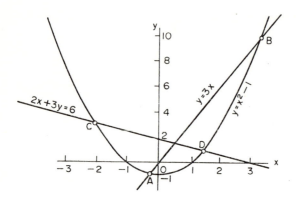

FIGURE 6.8

In each case, the points representing these ordered pairs may be plotted and will show the shapes of the graphs of the respective solution sets in $V \times V$. The graphs are shown above (Fig. 6.8).

In the case of (iii), $x^2 = 4$ is equivalent to $x = 2$ and $x = -2$. The solution set for $x = 2$ consists of the infinite set of ordered pairs, the first element of each ordered pair being 2. The graph of $x = 2$ is thus a straight line parallel to Oy at a distance $+2$ units from it. Similarly the graph of $x = -2$ is a straight line parallel to Oy at a distance -2 units from it.

The graph of the sentence $x^2 \leqslant 4$ consists of these two straight lines and that part of the plane which lies between them (Fig. 6.9).

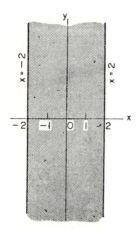

FIGURE 6.9

EXAMPLE 6.2

Show graphically the intersections of the solution sets of
(*a*) (i) and (iv) (*b*) (ii) and (iv) of Example 6.1.

(*a*) The intersection of the solution sets of (i) and (iv) will be represented by the points of intersection of the graphs of (i) and (iv), i.e. by A and B in Fig. 6.8.

Intersection set $= \{(-0.3, -0.9), (3.3, 9.9)\}$ approx.
$x = -0.3$, $y = -0.9$ and $x = 3.3$, $y = 9.9$ are the approximate solutions of the simultaneous equations $y = 3x$ and $y = x^2 - 1$.

(*b*) For (ii) and (iv), the intersection set consists of the two ordered pairs represented by C and D.

Intersection set $= \{(1.4, 1.1), (-2.1, 3.4)\}$ approx.
$x = 1.4$, $y = 1.1$ and $x = -2.1$, $y = 3.4$ are the approximate solutions of the simultaneous equations $2x + 3y = 6$ and $y = x^2 - 1$.

EXERCISES 6.2

1. Draw the graphs of the relations established by the following equations in the Cartesian set $A \times A$ where $A = \{0, 1, 2, 3, 4, 5\}$.
   (i) $y = 2x - 1$ (ii) $y < 2x - 1$ (iii) $y = 4 - x$ (iv) $y > 4 - x$
   (v) $x^2 + y^2 = 25$.

2. Draw graphs of the solution sets, in two dimensions, of the following open sentences in $V \times V$ where V is the set of real numbers.
   (i) $x = 2y$ (ii) $x - 2y < 0$ (iii) $a = 3b$ (*a* and *b* integers) (iv) $2x + y = 4$
   (v) $2x + y > 4$ (vi) $x^2 + y^2 = 25$ (vii) $x^2 + y^2 \leqslant 25$ (viii) $x = -3$
   (ix) $y = 1$ (x) $x^2 \leqslant 9$ (xi) $x^2 + y^2 = 0$ (xii) $x^2 + y^2 = -1$ (xiii) $xy = 8$
   (xiv) $xy = 0$ (xv) $xy > 0$.

3. With reference to Exercise 2, show graphically the intersections of the solution sets of
   (*a*) (i) and (iv) (*b*) (i) and (xiii) (*c*) (vi) and (x) (*d*) (vi) and (i).

4. Draw graphs to solve the following pairs of equations
   (*a*) $x^2 + y^2 = 4$ (*b*) $2x - y = 6$ (*c*) $x^2 + y^2 = 25$
   $\qquad y = x^2$ $\qquad 4x - 2y = 10$ $\qquad x + y = 8$
   Comment on the geometrical and algebraic significance of the graphs in (*b*) and (*c*).

6.7 Domain and Range of a Relation

Let R be a relation from the set A to the set B. Then R is a set of ordered pairs (x, y) such that $x \in A$, $y \in B$ and xRy.

DEFINITION 6.5

We define the *domain of the relation R* to be the subset of A which comprises all those elements x of A which belong to the elements (x, y) of R, i.e.

Domain $= \{x : x \in A$ and $(x, y) \in R\}$

Similarly, we define the *range of the relation R* to be the subset of B which comprises all those elements y of B which belong to elements (x, y) of R, i.e.

Range $= \{y : y \in B$ and $(x, y) \in R\}$

Let us take as examples the relations R_1 to R_7 which were considered in the last section. Their domains and ranges are tabulated below.

Relation	Domain	Range
R_1	$\{1, 2, 3\}$	$\{2, 3, 4\}$
R_2	$\{1, 2, 3, 4\}$	$\{1, 2, 3, 4\}$
R_3	{real numbers}	{real numbers}
R_4	{real numbers}	{real numbers}
R_5	$\{x : -3 \leqslant x \leqslant 3\}$	$\{y : -3 \leqslant y \leqslant 3\}$
R_6	{real numbers}	{positive real numbers and 0}
R_7	{positive real numbers and 0}	{real numbers}

The graphs of the relations R_1 and R_2 consist of a finite number of isolated points. The range and domain in these cases are said to be *discrete*. For the remaining relations, the graphs are smooth curves, with no gaps, consisting of an infinite number of points. In such cases, the domain and range are said to be *continuous*. (For R_5 the region enclosed by the circle has the same domain and range as the circle.)

6.8   *Function*

If we examine the graph of the relation R_2 (Fig. 6.4(*b*)), it is seen that for some values of x in the domain, more than one corresponding value of y is found in the range. On the other hand, on examining the graph of the relation R_3 (Fig. 6.5(*a*)), it is seen that for each x in the domain, there is one and only one y in the range. The relation R_3 establishes a *mapping* (see Section 6.5) from the set V into the set V where V is the set of real numbers. A relation such as R_3 is said to be a *function* whilst R_2 is not. A function is therefore a special kind of relation and is defined as follows.

DEFINITION 6.6
A function is a relation such that for every x in the domain of the relation, there is exactly one corresponding y in the range.

Graphically, this means that a relation R is a function if no straight line drawn through points of the domain, parallel to Oy, meets the graph of R in more than one point.

This criterion is satisfied in the case of the graph of R_6 (Fig. 6.10(a)) so

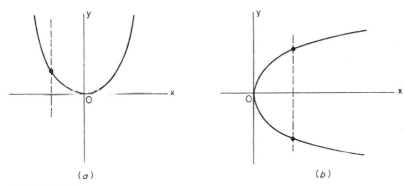

(a)   (b)

FIGURE 6.10
(a) Graph of $R_6 = \{(x, y):y = x^2\}$ (b) Graph of $R_7 = \{(x, y):y^2 = x\}$
    (a function).   (not a function).

that the relation R_6 is a function or more briefly $y = x^2$ is a function. On the other hand, this criterion is *not* satisfied in the case of the graph of R_7 (Fig. 6.10(b)) so that the relation R_7 is *not* a function, i.e. $y^2 = x$ is *not* a function.

6.9   *Functional Notation*

The symbols F, f, ϕ are commonly used to denote functions.

Each element of the domain of the function F corresponds to exactly one element y of its range. We denote this value of y, i.e. the second element of the ordered pair (x, y), by $F(x)$. $F(x)$, read "F at x", is "the value of the function at x". Thus y and $F(x)$ are two different symbols representing the same quantity and thus

$$y = F(x)$$

Unless it is explicitly stated otherwise, it will be assumed from now on that a function F is a relation in the Cartesian set $V \times V$ where V is the set of real numbers.

A function may be defined in various ways. It is usually defined by a formula or equation.

A *linear function* is the set of ordered pairs defined by an equation of the form

$$y = mx + b \qquad (m \neq 0)$$

which is of the first degree in x and y. The graph of any linear function is a *straight line* (see Fig. 6.5(a))

Consider the linear function

$$F = \{(x, y) : y = 2x - 1\}$$

The value of y when $x = 1$ is

$$F(1) = 2(1) - 1 = 1$$

and

$$F(0) = 2(0) - 1 = -1$$
$$F(-1) = 2(-1) - 1 = -3$$
$$F(\tfrac{1}{2}) = 2(\tfrac{1}{2}) - 1 = 0$$

The *quadratic function* is defined by an equation of the form

$$y = ax^2 + bx + c \qquad (a \neq 0)$$

The graph of any quadratic function is a *parabola* (see Fig. 6.7). Consider the quadratic function

$$F = \{(x, y) : y = x^2 - 5x + 6 = (x - 2)(x - 3)\}$$

then we have

$$F(1) = 1^2 - 5.1 + 6 = 2$$
$$F(0) = 0 - 5.0 + 6 = 6$$
$$F(2) = F(3) = 0$$
$$F(a) = a^2 - 5a + 6$$
$$F(-x) = (-x)^2 - 5(-x) + 6 = x^2 + 5x + 6$$

If a function is defined by an equation $y = F(x)$, it is customary to refer to the equation as the *function $y = F(x)$*.

A function may also be defined by a table, i.e. by a tabulated set of ordered pairs, thus:

x	−2	−1	0	1	2	3	4	5	6	7
y	5	0	−3	−4	−3	0	5	12	21	32

The domain and range of this function are respectively

$$\{-2, -1, 0, 1, 2, 3, 4, 5, 6, 7\} \quad \text{and} \quad \{-4, -3, 0, 5, 12, 21, 32\}$$

The graph of this function consists of a set of points lying on a parabola (Fig. 6.11).

A function may also be defined by a graph which displays the points representing the set of ordered pairs which constitutes the function. The graph in Fig. 6.11 would therefore serve to define the same function as the table above.

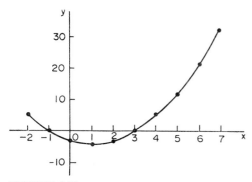

FIGURE 6.11

The graph in Fig. 6.10(a) would similarly serve to define the function $y = x^2$.

6.10 Inverses of Relations and Functions

The graph of the relation

$$R = \{(1, 2), (2, 4), (3, 4), (4, 5), (5, 5)\}$$

consists of the points heavily marked in Fig. 6.12. Its domain is $\{1, 2, 3, 4, 5\}$ and its range is $\{2, 4, 5\}$.

Consider the relation

$$R' = \{(2, 1), (4, 2), (4, 3), (5, 4), (5, 5)\}$$

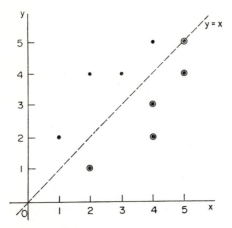

FIGURE 6.12

which is derived from R by interchanging the coordinates of each ordered pair in R. The relation R' is said to be the *inverse* of the relation R. The graph of R' consists of the five points, marked with open circles in Fig. 6.12, and is the image of the graph of R in the line $y = x$.

DEFINITION 6.7
The *inverse* of a relation is the relation obtained by interchanging the co-ordinates of all the ordered pairs which constitute the given relation.

Thus any ordered pair (a, b) of the given relation becomes the ordered pair (b, a) of the inverse.

It is easy to prove that the point $P'(b, a)$ is the mirror image of the point $P(a, b)$ in the line $y = x$ (Fig. 6.13).

Let PN and P'N' be perpendiculars drawn from P and P' respectively to the axes Ox and Oy respectively. The triangles OPN and OP'N' are obviously congruent. Therefore

$$\angle PON = \angle P'ON' = \alpha \text{ (say)}$$
$$OP = OP'$$

Since the line $y = x$ bisects the angle between the axes of coordinates, it follows that

$$\angle P'OM = \angle POM = 45° - \alpha$$

Therefore the triangles POM, P'OM are congruent.
Therefore OM is the perpendicular bisector of PP'.
Therefore $P'(b, a)$ is the mirror image of $P(a, b)$ in the line $y = x$.
It follows that the graph of the inverse of a relation is the reflection of the graph of that relation in the line $y = x$.

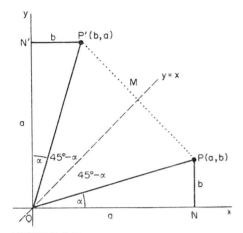

FIGURE 6.13

Since a function is a relation, its inverse is defined by Definition 6.7. The inverse of a function may be another function or it may be a relation which is *not* a function. If a function is defined by an equation, then the equation of its inverse is obtained by interchanging the rôles of x and y in the given equation.

Consider the function $F = \{(x, y):y = x^2\}$.

In Fig. 6.14(a), the graph of the function F, defined by the equation $y = x^2$, is shown. The graph of its inverse F' will be obtained by reflecting

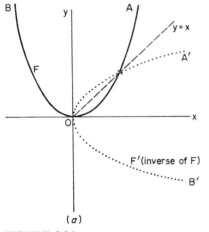

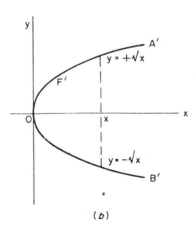

(*a*)   (*b*)

FIGURE 6.14

(*a*) $F = \{(x, y):y = x^2\}$ (*a function*).   (*b*) $F' = \{(x, y):y^2 = x, \text{ or } y = \pm \sqrt{x}\}$ (*not a function*).

the graph of F in the line $y = x$. The graph of F' is shown in Fig. 6.14(b) and, from this graph, it is clear that *two* values of y in the range of F' correspond to each value of x in its domain.

Thus F' is a relation which is not a function.

The equation which defines F' is $y^2 = x$ or $y = \pm\sqrt{x}$ which is derived from the equation $y = x^2$ which defines F by interchanging x and y.

Thus $F' = \{(x, y) : y^2 = x \text{ or } y = \pm\sqrt{x}\}$.

If we now restrict the domain of F so that $x \geqslant 0$, the graph of F will be restricted to the portion OA (Fig. 6.14(a)) and its inverse will be OA' (Fig. 6.14(b)) which is the graph of a *function* $\{(x, y) : y = +\sqrt{x}\}$. Similarly OB' is the graph of the inverse of OB and OB' is the graph of the *function* $\{(x, y) : y = -\sqrt{x}\}$.

EXAMPLE 6.3

Sketch the graphs of the following functions and their inverses in $V \times V$:

(a) $2y = x + 4$ (b) $3x + y = 6$ (c) $y = -2$ (d) $y = x^2 - 4$

What are the equations of the inverses and which of them are functions? The graphs of (a) and (b) are straight lines which may be sketched at once if two points on each are known.

(a) $2y = x + 4$. Two points on the graph of this function are obviously $(0, 2)$ and $(-4, 0)$.

The inverse is the *function* $2x = y + 4$. Its graph is the reflection of the graph of $2y = x + 4$ in the line $y = x$ (Fig. 6.15(a)).

(b) $3x + y = 6$

Its inverse is the *function* $3y + x = 6$ (Fig. 6.15(b)).

(c) The inverse of the function $y = -2$ is $x = -2$ which is *not* a function (Fig. 6.15(d)).

(d) The graph of $y = x^2 - 4$ is a parabola, and the points $(2, 0)$, $(-2, 0)$ and $(0, -4)$ are obviously on it.

If necessary, other corresponding pairs of values of x and y may be tabulated and plotted. The shape and position of the graph of $y = x^2 - 4$ is as illustrated in Fig. 6.15(d).

This graph shows that for every x in the domain there is one and only one y in the range, so that $y = x^2 - 4$ is a *function*.

The equation of the inverse is $x = y^2 - 4$ or $y^2 = x + 4$. The graph of the inverse is the reflection of the graph of $y = x^2 - 4$ in the line $y = x$. This is obviously an equal parabola as illustrated in Fig. 6.15(d).

The domain of the relation defined by $y^2 = x + 4$ is $\{x : x \geqslant -4\}$ and its range is $\{$real numbers$\}$. There is a two-to-one correspondence between the domain and the range. Thus $y^2 = x + 4$ is *not* a function.

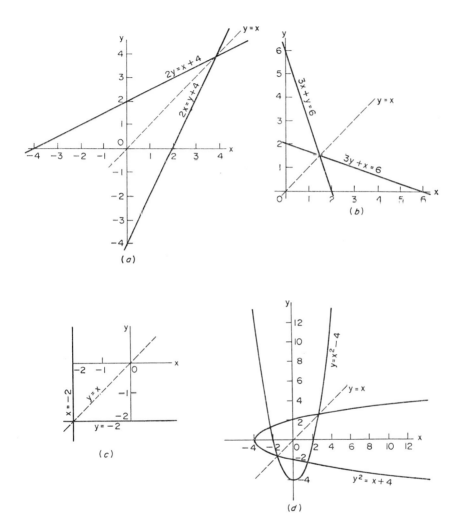

FIGURE 6.15

EXAMPLE 6.4

For $F = \{(x, y) : y = x^2 + 3x - 4\}$

(a) find $F(0)$, $F(1)$, $F(-4)$, $F(2x)$, $F(-x)$, $F(1/x)$, $F(x + a) - F(x - a)$

(b) sketch the graph of $y = x^2 + 3x - 4$ in $V \times V$

(c) find the domain and range of F in $V \times V$.

(a)

$$F(0) = -4$$

$$F(1) = F(-4) = 0$$

$$F(2x) = (2x)^2 + 3(2x) - 4 = 4x^2 + 6x - 4$$

$$F(-x) = (-x)^2 + 3(-x) - 4 = x^2 - 3x - 4$$

$$F\left(\frac{1}{x}\right) = \left(\frac{1}{x}\right)^2 + 3\left(\frac{1}{x}\right) - 4 = \frac{1}{x^2} + \frac{3}{x} - 4$$

$$F(x + a) - F(x - a)$$
$$= [(x + a)^2 + 3(x + a) - 4] - [(x - a)^2 + 3(x - a) - 4]$$
$$= [(x + a)^2 - (x - a)^2] + 3[(x + a) - (x - a)]$$
$$= 4ax + 6a$$

(b) Tabulate corresponding pairs of values of x and y.

x	-5	-4	-3	-2	-1	0	1	2
$y = (x - 1)(x + 4)$	6	0	-4	-6	-6	-4	0	6

The graph of $y = x^2 + 3x - 4$ is shown in Fig. 6.16.

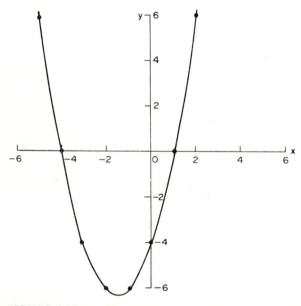

FIGURE 6.16

(c) The domain of this function is {real numbers}.
By symmetry, the minimum value of y corresponds to $x = -1\frac{1}{2}$ and is therefore

$$(-1\tfrac{1}{2} - 1)(-1\tfrac{1}{2} + 4) = -\frac{5}{2} \times \frac{5}{2} = -6\tfrac{1}{4}$$

The range of this function is therefore $\{y:y \geqslant -6\tfrac{1}{4}\}$.

EXERCISES 6.3

In the following exercises let V be the set of real numbers.

1. For the functions described below, find the domain and range and establish the correspondence by means of a formula or a table.
    (a) Mary, Jane and Elizabeth are married to John, Fred and Tom respectively.
    (b) To each integer, there corresponds a number equal to half its value.
    (c) To each real number, there corresponds its negative.

2. The graphs in Fig. 6.17 represent relations in $V \times V$. Which of these relations are functions?

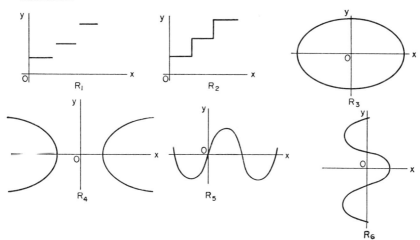

FIGURE 6.17

3. Graph the following sets:
    (a) $\{(1, 2), (1, 3), (2, 3), (3, 4)\}$
    (b) $\{(1, 2), (2, 3), (3, 4)\}$
    (c) $\{(1, 2), (2, 4), (3, 4), (4, 4)\}$
    Are these sets functions or relations?
    Graph their inverses. Are the inverses functions or relations?

4. Sketch the graphs of the following functions and their inverses in $V \times V$:
    (a) $y = 2x$ (b) $y = 2x + 3$ (c) $3x + 2y = 6$ (d) $y = 2$ (e) $xy = 4$
    (f) $y = 4 - x^2$.
    What are the equations of the inverses and which are functions?

5. Find the ranges of the following functions corresponding to the domains stated:
   (a) $f(x) = x - 1$, $x = 0, 1, 2$ (b) $g(x) = -x$, $x > 0$
   (c) $\phi(x) = |x|$, x is any real number   (d) $F(x) = (x - 1)^2$, x is any real
   number   (e) $f(x) = (x - 1)^{-1}$, $x > 1$ (f) $f(u) = (1 - u^2)$, $0 \leqslant u \leqslant 1$

   (g) $\phi(v) = \dfrac{1}{1 + v^{-1}}$, $v > 0$

6. Find the domain of x in V so that the ranges of the following functions are also
   in V.

   (a) $f(x) = \sqrt{(x - 1)}$ (b) $f(x) = \sqrt[3]{x^2}$ (c) $f(x) = \sqrt{(x^2 - 2x + 1)}$
   (d) $f(x) = x + x^{1/2} + x^{1/4}$ (e) $f(x) = (-x)^{1/2}$

7. If $f(x) = x^2 - 2x + 1$, find
   (a) $f(1)$ (b) $f(-2)$ (c) $f(3x)$ (d) $f(x + a) - f(x)$ (e) $f(-x)$

   (f) $f\left(x + \dfrac{1}{x}\right)$.

8. $F = \{(x, y): y = x^2 + 2x - 3\}$ is a function in $V \times V$.

   (a) Find $F(0)$, $F(1)$, $F(-3)$, $F(\frac{1}{2})$, $F\left(\dfrac{1}{x}\right)$, $F(3x)$, $F\left(1 - \dfrac{1}{x}\right)$.

   (b) Draw the graph of the function F.
   (c) Find the domain and range of F.

7 Linear programming

7.1 Linear Functions

Reference has already been made to linear functions in Section 6.9. We now consider these functions in more detail.

DEFINITION 7.1
A *linear function* is a set of ordered pairs (x, y) in $V \times V$ defined by the equation

$$y = mx + b \qquad m \neq 0$$

i.e.

$$F = \{(x, y) : y = mx + b\}$$

It is customary to talk of the "linear function $y = mx + b$" as an abbreviation for the "linear function defined by $y = mx + b$" and of the "graph of $y = mx + b$" when we really mean the "graph of the linear function defined by $y = mx + b$". We also refer to $y = mx + b$ as the "equation of a linear function".

7.2 The Graph of a Linear Function

Let $P_1(x_1, y_1)$ and $P_2(x_2, y_2)$ be any two points on the graph of $y = mx + b$ (Fig. 7.1).

Let P_2N and P_1N be respectively parallel to Ox and Oy.

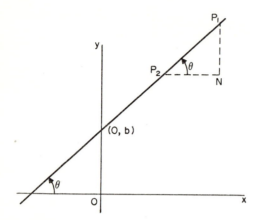

FIGURE 7.1   *Graph of y = mx + b.*

From the right-angled triangle P_1NP_2, the slope of the straight line P_1P_2 is

$$\tan \theta = \frac{NP_1}{P_2N} = \frac{y_1 - y_2}{x_1 - x_2}$$

But $y_1 = mx_1 + b$ and $y_2 = mx_2 + b$. Therefore

$$y_1 - y_2 = m(x_1 - x_2)$$
$$m = \frac{y_1 - y_2}{x_1 - x_2}$$

Therefore the slope of the straight line P_1P_2 is m.

It follows that the straight lines joining *every* pair of points on the graph of $y = mx + b$ have a constant slope m, i.e. have the same direction. This can only mean that the points representing all ordered pairs belonging to F lie on one straight line whose slope is m.

Each ordered pair belonging to the set defined by the linear equation $y = mx + b$ corresponds to a point lying on a unique straight line whose equation is $y = mx + b$. There is thus a one-to-one correspondence between a linear equation and a straight line.

It is obvious that $(0, b) \in F$, so that the graph of $y = mx + b$, makes an intercept b on the y-axis (Fig. 7.1).

The graph of the linear function $y = \frac{1}{2}x - 3$ has a slope $1/2$ and makes an intercept -3 on the y-axis (Fig. 7.2(*a*)).

The straight lines $y = \frac{1}{2}x + b$ ($b = -3, -2, -1, 0, 1, 2$) are parallel and make intercepts $-3, -2, -1, 0, 1, 2$ on Oy (Fig. 7.2(*a*)). The straight lines

$y = mx + 2$ for various values of the slope m comprise a set of lines with a
common intercept $+2$ on Oy (Fig. 7.2(b)).

Let (x_1, y_1) be a point on the straight line $y = mx + b$. Thus

$$y_1 = mx_1 + b$$
$$b = y_1 - mx_1$$

so that $y = mx + y_1 - mx_1$ and therefore

$$y - y_1 = m(x - x_1)$$

This is the equation of the straight line of slope m through the point (x_1, y_1).

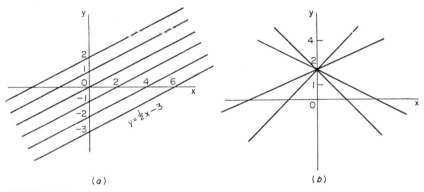

(a)   (b)

FIGURE 7.2
(a) Graphs of $y = \frac{1}{2}x + b$.   (b) Graphs of $y = mx + 2$.

7.3 Graphs of Linear Inequalities

The points which represent the set of ordered pairs defined by the equation
$ax + by + c = 0$ lie on a straight line.

We enquire, what are the graphs of $ax + by + c \geq 0$?

Let us consider the linear function in the form $y = mx + b$. What is the
graph of the linear inequality $y < mx + b$?

Let $P(x_0, y_0)$ be any point on the straight line $y = mx + b$ (Fig. 7.3) and
let $Q(x_0, y_1)$ be any point below this line on the ordinate PN of the point P.
We have

$$y_0 = mx_0 + b$$

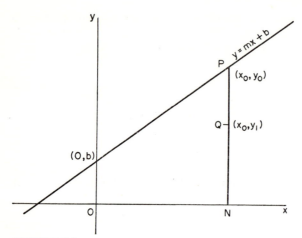

FIGURE 7.3

and

$$y_1 < y_0$$

Therefore

$$y_1 < mx_0 + b$$

Thus the coordinates of any point *below* the line $y = mx + b$ satisfy the inequality $y < mx + b$.

Similarly, the coordinates of any point *above* the line $y = mx + b$ satisfy the inequality $y > mx + b$.

It follows that the solution set of the linear inequality $y < mx + b$ is the set of ordered pairs represented by all points *below* the line $y = mx + b$ and that of $y > mx + b$ is the set of ordered pairs represented by all points *above* the line $y = mx + b$.

The line $y = mx + b$ divides the plane into two "open half planes" of which the lower half plane is the graph of $y < mx + b$ and the upper half plane is the graph of $y > mx + b$. The word "open" implies that all ordered pairs represented by points *on* the line $y = mx + b$ are excluded from the solution sets of the linear inequalities.

EXAMPLE 7.1

(A) Draw the graph of $2y + 3x + 6 < 0$.

Express the inequality in the equivalent form $y < -\frac{3}{2}x - 3$. First draw the graph of $y = -\frac{3}{2}x - 3$ (Fig. 7.4). The graph of the solution set of the

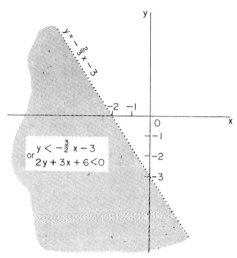

FIGURE 7.4

inequality $y < -\frac{3}{2}x - 3$ or $2y + 3x + 6 < 0$ is the half plane (shaded) below the straight line.

(B) Draw the graph of $x \geqslant -4$.

In this case, the graph of the straight line $x = -4$ is parallel to Oy and distant -4 units from it. The graph of $x \geqslant -4$ is the *closed* half plane to the right of the straight line $x = -4$ and the word "closed" implies that all points *on* the line $x = -4$ belong to the graph of $x \geqslant -4$ (Fig. 7.5).

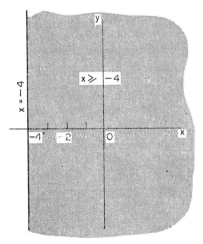

FIGURE 7.5

7.4 *Simultaneous Linear Inequalities*

In various practical situations, the variables x and y frequently must satisfy two or more inequalities simultaneously. In cases where two inequalities are satisfied simultaneously, the corresponding straight line graphs (assuming that they are not parallel) will divide the plane into 4 regions. In *one* of these regions, the coordinates of all points will satisfy the two inequalities simultaneously.

EXAMPLE 7.2
Show the region of 2-dimensional coordinate space which is the graph of the solution set of the inequalities

$$2x + 3y - 12 > 0 \qquad 2y - 3x - 20 < 0$$

when $x \geqslant 0$ and $y \geqslant 0$.
    These two inequalities may be written in the equivalent forms

$$y > -\frac{2}{3}x + 4 \qquad y < \frac{3}{2}x + 10$$

    In Fig. 7.6, line L_1 is the graph of $y = -\frac{2}{3}x + 4$. Therefore the graph of the inequality

$$y > -\frac{2}{3}x + 4$$

is the half plane *above* line L_1.
    Line L_2 is the graph of $y = \frac{3}{2}x + 10$. Therefore the graph of the inequality

$$y < \frac{3}{2}x + 10$$

is the half plane *below* line L_2.
    Both inequalities therefore hold for the coordinates of all points of the intersection of these two half planes. Since x and y are each to be positive, the graph of the solution set is the shaded space (Fig. 7.6).
    Using algebraic methods, we have

$$2x + 3y - 12 > 0 \qquad 4x + 6y - 24 > 0$$
$$3x - 2y + 20 > 0 \qquad 9x - 6y + 60 > 0$$

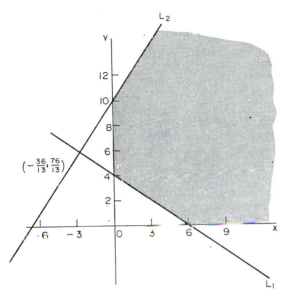

FIGURE 7.6

Since $a > 0, b > 0 \Rightarrow a + b > 0$, we may add these inequalities to eliminate y:

$$13x + 36 > 0$$

$$x > -\frac{36}{13}$$

This implies that the inequalities can only be satisfied simultaneously for values of x exceeding $-36/13$.

[The point of intersection of lines L_1 and L_2 may be found by solving their equations simultaneously. It is the point $(-36/13, 76/13)$.]

On the other hand, nothing can be discovered about y if we try to eliminate x. We have

$$6x + 9y - 36 > 0$$
$$6x - 4y + 40 > 0$$

It is not valid to subtract these inequalities to eliminate x since $a > 0, b > 0$ gives no information about the sign of $a - b$. The graphical method shows that y may take any value.

The graphical method is usually the most convenient method for finding the solution sets of simultaneous linear inequalities.

EXAMPLE 7.3
Show by shading the regions of the (x, y) plane which are the graphs of

    (a) $x \geqslant 0, y \geqslant 0, 5x + 2y \leqslant 20$

    (b) $x \geqslant 0, y \geqslant 0, 5x + 2y \geqslant 20$

    (c) $x \geqslant 0, y \geqslant 0, 2x + 3y \leqslant 12, 5x + 2y \leqslant 20.$

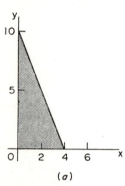
(a)

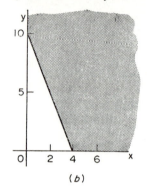
(b)

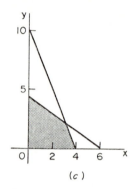
(c)

FIGURE 7.7

For each of the cases (a), (b), and (c), the conditions $x \geqslant 0, y \geqslant 0$ restrict all graphs to the first quadrant of the plane including all points on the positive segments of the coordinate axes (Fig. 7.7).

In (a) the graph comprises the triangular area below the line $5x + 2y = 20$ bounded by it and the axes of coordinates.

In (b) the graph comprises the area above the line $5x + 2y = 20$ in the first quadrant, and bounded by the axes themselves.

In (c) the graph comprises the quadrilateral common to the areas under both lines in the first quadrant, and bounded by the axes.

EXERCISES 7.1

1. Draw graphs of the following inequalities
       (a) $y < -3$ (b) $x \geqslant 2$ (c) $y > 3x - 2$
       (d) $y < 5 - 2x$ (e) $2x - 5y + 15 > 0$

2. Show, by shading, the graphs of the following simultaneous inequalities:
       (a) $y \geqslant 2x - 3, x \geqslant 3$
       (b) $2x + 3y \geqslant 6, 2x - y + 1 \geqslant 0$
       (c) $x \geqslant 0, y \geqslant 0, 2x + 3y - 6 \geqslant 0$
       (d) $x \geqslant 0, y \geqslant 0, 2x + 3y - 6 \leqslant 0$
       (e) $x \geqslant 0, y \geqslant 0, 2x + 3y \leqslant 6, 5x + 2y \leqslant 10$
       (f) $y \leqslant 3, x \leqslant 4, 4x + 5y - 20 \geqslant 0$

3. Shade the regions of the plane which contain the points whose coordinates satisfy the simultaneous inequalities.
       (a) $x^2 + y^2 < 16, y > x + 2$ (b) $y > x^2 - 1, 2y + x + 1 < 0$

7.5 Maximizing and Minimizing

The graphs of the linear functions $3y + 2x = 6,\ 12,\ 18$ are parallel lines (Fig. 7.8) crossing the first quadrant of the Cartesian plane. We note that the higher the constant term, the farther is the corresponding line from the origin. (It is assumed that the constant terms are all positive.)

At all points on any one of these lines, $(3y + 2x)$ has a constant value, say $c > 0$ which increases as the distance from the origin of the line selected increases.

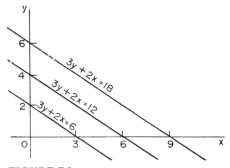

FIGURE 7.8

Now suppose we require the maximum (or minimum) value of $(3y + 2x)$ subject to specified restrictive conditions on the values of x and y. In practical situations, these restrictive conditions are generally expressed by one or more linear inequalities which x and y satisfy.

If, as is frequently the case, these conditions restrict the values of x and y to the coordinates of points in the first quadrant, then the maximum (or minimum) value of $(3y + 2x)$ is the value of $c > 0$ corresponding to the line $3y + 2x = c$ which may be drawn farthest from (or nearest to) the origin under these restrictive conditions.

In more general terms, let us consider the problem of finding the maximum (minimum) of $(ax + by)$ subject to certain restrictive conditions.

The graphs of the linear functions

$$ax + by = c \quad \text{or} \quad y = -\frac{a}{b}x + \frac{c}{b}$$

for varying values of c will comprise a set of parallel lines.

(I) If $a > 0,\ b > 0$, the intercepts c/b on the y-axis are positive or negative according as c is positive or negative and their magnitudes are proportional to

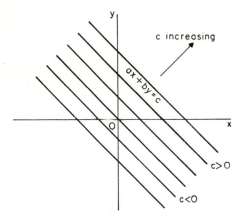

FIGURE 7.9(a)   Graph of $ax + by = c$.
$a > 0, b > 0$

the values of c. It follows that as c increases, the corresponding line will be found farther to the right (Fig. 7.9(a)).

If $a < 0$, $b < 0$, we have a similar set of parallel lines, but as c increases the corresponding line will be found farther to the left.

(II) If $a > 0$, $b < 0$, we obtain a set of parallel lines as in Fig. 7.9(b).

The intercepts c/b are negative or positive according as c is positive or negative and their magnitudes are proportional to c.

It follows that as c increases, the corresponding line will be found farther to the right.

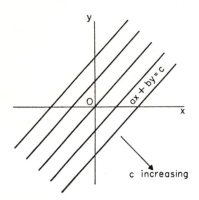

FIGURE 7.9(b)   Graph of $ax + by = c$.
$a > 0, b < 0$

If $a < 0$, $b > 0$, we have a similar set of parallel lines, but as c increases the corresponding line will be found farther to the left.

The maximum (minimum) of a linear function $(ax + by)$, subject to specified restrictions on the values of x and y, is the value of c which corresponds to the particular line of the set $ax + by = c$ which may be drawn with the greatest (least) value of c through a point or points of the region of the Cartesian plane defined by the restrictive conditions.

The name *Linear Programming* has been given to the solution of problems of this nature. Some illustrative examples of linear programming follow.

EXAMPLE 7.4

Subject to the restrictions $x \geqslant 0$, $y \geqslant 0$, $5x + 3y \leqslant 30$, $4x + 9y \leqslant 36$ what is the maximum value of

(a) $x + y$ (b) $3y + 2x$ (c) $x + 9y$?

What is the maximum value in each case if x and y may take only integral values

The restrictions $x \geqslant 0$, $y \geqslant 0$ mean that we are concerned only with points in the first quadrant.

Draw the graphs of $5x + 3y = 30$ and $4x + 9y = 36$ (Fig. 7.10). By solving these equations simultaneously, it will be found that their graphs intersect at the point B $(54/11, 20/11)$.

The graph of the simultaneous inequalities

$$x \geqslant 0 \qquad y \geqslant 0 \qquad 5x + 3y \leqslant 30 \qquad 4x + 9y \leqslant 36$$

is the interior and the four sides of the quadrilateral OABC.

(a) We require the maximum of $(x + y)$ for the number pairs corresponding to those points which lie in or on the quadrilateral OABC. The line $x + y = 2$ has been drawn in Fig. 7.10. We now draw a line, parallel to $x + y = 2$ passing through a point or points of the region OABC at maximum distance from the origin. The value of $(x + y)$ at any point on this line will be the maximum of $(x + y)$.

In this case, the line at maximum distance from the origin passes through the point B$(54/11, 20/11)$. Therefore

$$(x + y)_{max} = \left(\frac{54}{11} + \frac{20}{11}\right) = \frac{74}{11} = 6\frac{8}{11}$$

(b) The line parallel to $3y + 2x = 6$ at maximum distance from the origin again passes through the point B. Therefore

$$(3y + 2x)_{max} = \left(3 \times \frac{20}{11} + 2 \times \frac{54}{11}\right) = \frac{168}{11} = 15\frac{3}{11}$$

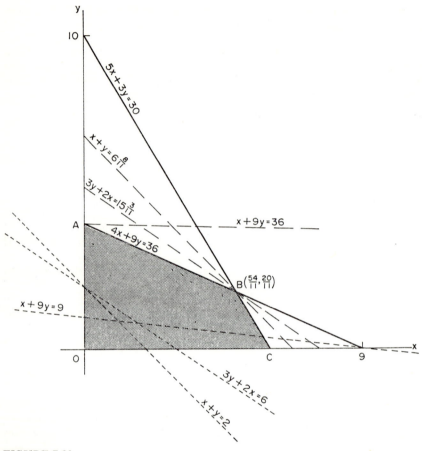

FIGURE 7.10

(c) The line at maximum distance from the origin which is parallel to
$x + 9y = 9$, is the line which passes through $A(0, 4)$. Therefore

$$(x + 9y)_{max} = (0 + 9 \times 4) = 36$$

In cases where x and y may take integral values only, we shall be concerned
only with those ordered pairs which are represented by the lattice points
having integral coordinates within or on the sides of the quadrilateral OABC.
(a) The line parallel to $x + y = 2$ which is at maximum distance from the
    origin passes through $(5, 1)$, $(4, 2)$, $(6, 0)$. Thus $(x + y)_{max}$ occurs for
    each of these ordered pairs and is 6.

(b) The line parallel to $3y + 2x = 6$, which is at maximum distance from the origin passes through the lattice point $(4, 2)$. Thus

$$(3y + 2x)_{max} = (3 \times 2 + 2 \times 4) = 14$$

(c) Since the point $A(0, 4)$ is a lattice point of the region, $(x + 9y)_{max} = 36$ as before.

Some typical practical problems of Linear Programming are illustrated by Examples 7.5, 7.6 and 7.7.

EXAMPLE 7.5

A farmer intends to plant his 18 acre field with potatoes and beans. The cost of seed, fertilizers, etc., for potatoes is £4 per acre and for beans £3 per acre. The farmer has a capital of £60 and makes a profit of £6 per acre on potatoes and £5 per acre on beans. How should he allocate the land for maximum profit?

	Potatoes	Beans
Cost per acre	£4	£3
Profit per acre	£6	£5
Let the number of acres planted be	x	y

Total profit = £$(6x + 5y)$
Total cost of planting = £$(4x + 3y) \leqslant$ £60
No. of acres planted = $x + y \leqslant 18$

In this problem, we therefore require the maximum of $(6x + 5y)$ subject to the following conditions:

$$x \geqslant 0, \quad y \geqslant 0$$
$$x + y \leqslant 18$$
$$4x + 3y \leqslant 60$$

With reference to Fig. 7.11 the graph of these simultaneous inequalities is the interior and 4 sides of the quadrilateral OABC. B is the point $(6, 12)$.

The line parallel to $6x + 5y = $ constant which is farthest from the origin is the line which passes through the point $B(6, 12)$. Therefore

$$(6x + 5y)_{max} = 6 \times 6 + 5 \times 12 = 96$$

The farmer must therefore plant 6 acres with potatoes and 12 acres with beans to make a maximum profit which will be £96.

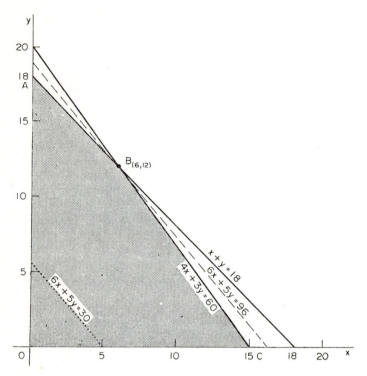

FIGURE 7.11

EXAMPLE 7.6

A businessman wishes to stock at least 60 units of a certain item. He can obtain them from two suppliers A and B. The cost of transport is $37\frac{1}{2}$p per unit from A and 25p per unit from B. He does not wish to pay more than £19 for transport. The cost per unit from A is £9 and from B is £12. Orders have to be made in dozens. What must he order from A and B to minimize his cost and what is this?

	Supplier A	*Supplier B*
Transport charges per unit	37.5p = £0.375	25p = £0.25
Cost price per unit	£9	£12
Let the number of units ordered be	x	y

Total number of units $= x + y \geqslant 60$
Transport charge $= £(0.375x + 0.25y) \leqslant £19$
Cost of goods $= £(9x + 12y) = 3 \times £(3x + 4y)$

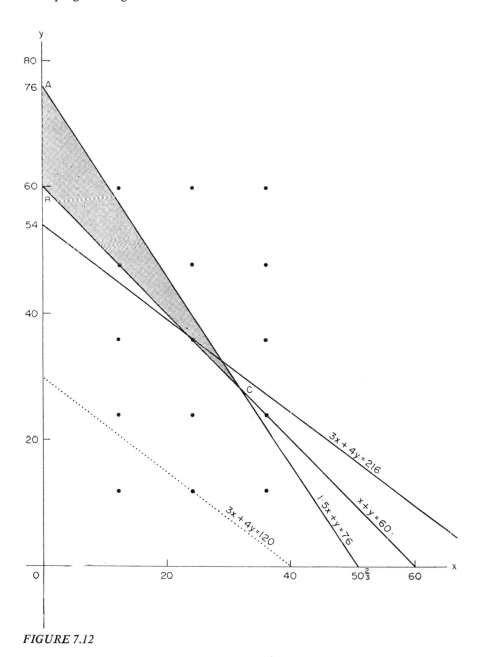

FIGURE 7.12

In this problem therefore we require the minimum of $3x + 4y$ subject to

$$x \geqslant 0 \qquad y \geqslant 0 \qquad x + y \geqslant 60$$

$$(0.375x + 0.25y) \leqslant 19$$

The last inequality may be simplified to $1.5x + y \leqslant 76$.

The graph of these inequalities is the triangle ABC (shaded) in Fig. 7.12. Since orders are taken in dozens only, we are interested only in values of x and y which are integral multiples of 12. In Fig. 7.12 lattice points are marked whose coordinates are each multiples of 12.

The line $3x + 4y =$ constant, which passes through a lattice point in the region ABC and is nearest to the origin, is the line which passes through the point (24, 36). On this line

$$3x + 4y = 3 \times 24 + 4 \times 36 = 216$$
$$(9x + 12y)_{min} = 3 \times 216 = 648$$

The business man must therefore buy 2 dozen units from A and 3 dozen units from B in order to minimize his cost. The minimum cost is £648.

It is clear from Fig. 7.12 that if the businessman were prepared to spend more on transport, the triangle ABC would become larger and include more lattice points through some of which a line $3x + 4y =$ constant may be drawn nearer to the origin. This would reduce the minimum cost of goods.

EXAMPLE 7.7

Two brands of tea are each mixed from three different grades of tea, A, B and C. A 1 lb packet of the first brand contains 3 oz of A, 6 oz of B and 7 oz of C, and yields a profit of 12.5p per packet. A 1 lb packet of the second brand contains 6 oz of A, 3 oz of B and 7 oz of C, and yields a profit of 7.5p per packet.

The amounts of tea available are 75 000 oz of A, 60 000 oz of B and 98 000 oz of C. How many packets of each brand should be made for maximum profit and what is this?

	A	B	C
Brand 1: Each packet contains	3 oz	6 oz	7 oz
Brand 2: Each packet contains	6 oz	3 oz	7 oz
Total number of ounces available	75 000	60 000	98 000

Brand 1: Profit per packet = 12.5p
Brand 2: Profit per packet = 7.5p

Let the number of packets of Brand 1 be x.
Let the number of packets of Brand 2 be y.

No. of ounces of grade A tea required $= 3x + 6y \leqslant 75\ 000$

$x + 2y \leqslant 25\ 000$

No. of ounces of grade B tea required $= 6x + 3y \leqslant 60\ 000$

$2x + y \leqslant 20\ 000$

No. of ounces of grade C tea required $= 7x + 7y \leqslant 98\ 000$

$x + y \leqslant 14\ 000$

$x \geqslant 0, \quad y \geqslant 0$

Subject to the above restrictions on x and y, we require to maximize the total profit

$$£(0.125x + 0.075y) = £(2.5x + 1.5y)/20$$

We therefore require to maximize $(2.5x + 1.5y)$.

The graph of the inequalities is the shaded region OABCD (Fig. 7.13).

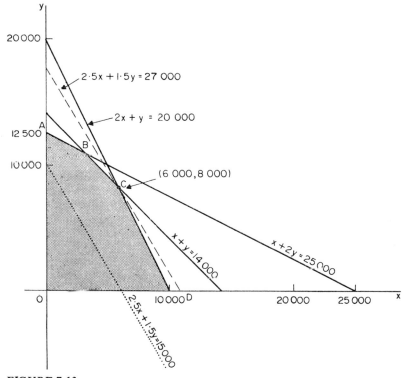

FIGURE 7.13

The line parallel to $2.5x + 1.5y = $ constant which is at maximum distance from the origin for points of the region is the line through the point C (6 000, 8 000).

$$\text{Maximum profit} = \pounds(2.5x + 1.5y)/20$$
$$= \pounds(2.5 \times 6\ 000 + 1.5 \times 8\ 000)/20$$
$$= \pounds1\ 350$$

To obtain this profit, 6 000 packets of brand 1 and 8 000 packets of brand 2 must be made.

EXERCISES 7.2

1. Find the maxima and minima, when they exist, of the function $(x + y)$ subject to the restrictions of no. 2 of Exercises 7.1. What will they be if x and y can only take integral values?

2. Show, on a graph, the region for which
   $$x \geqslant 0, \quad y \geqslant 0, \quad x + 4y \geqslant 8, \quad 6x + 5y \geqslant 30, \quad 9x + y \geqslant 9$$
   (a) With these restrictions, minimize
   (i) $3x + y$ (ii) $4x + 5y$ (iii) $x + 7y$
   stating the coordinates of the points giving these minimum values.
   (b) Repeat (a) if x and y may take only integral values
   $[6\frac{9}{13}, 21\frac{11}{19}, 8; 8, 25, 8]$

3. The coordinates (x, y) of a set of points satisfy the conditions
   $$x \geqslant 0, \quad 5 \geqslant y \geqslant 0, \quad y \leqslant 2x + 1, \quad x + y \leqslant 15$$
   Find the points of this set for which the functions
   (a) $3x + 5y - 2$ (b) $7x - 3y + 5$
   each have maximum and minimum values and what are they?
   $[(a)$ $(0, 0)$, min. -2; $(10, 5)$, max. 53
   (b) $(0, 1)$, min. $+2$; $(15, 0)$, max. $110)]$

4. A lawyer is paid £60 for each criminal case, £30 for each civil case and £25 for each matrimonial case. He takes precisely 250 cases per year, of which at least 75 are matrimonial cases. He does not wish to take more than 100 criminal cases per year. His annual expenses are £3 000. How many of each type of case should he take to obtain a maximum net income and what is this income?
   [75, 75, 100; £10 125]

5. A dealer can buy a certain item from two suppliers A and B. He requires at least 600 units of this item. The transport available at A and B is not capable of carrying more than 500 units and 300 units respectively. The cost of delivery from A is £1.50 per unit plus £300 and from B is £1 per unit plus £400. The cost price per unit from supplier A is £6 if at least 100 units are purchased and from supplier B is £4 if at least 150 units are purchased. The dealer will not pay more than £1 500 for transport but he agrees to purchase at least 100 units from A and 150 units from B.
   What must he order from A and B to minimize his cost and what is this?
   [300 units from each of A and B; minimum cost £4 450]

6. Two alloys A and B are made by mixing copper and zinc in different proportions. A contains 3 parts of copper to 2 parts of zinc and B contains 1 part of copper to 2 parts of zinc. The profits obtained per ton on the alloys A and B are £5 and £4 respectively. 90 tons of copper and 80 tons of zinc are available.

How many tons of each alloy should be made to obtain maximum profit and what is the profit? If the maximum capacity of the factory is 155 tons what are the tonnages and the maximum profit?

[125, 45; £805; $143\frac{3}{4}$, $11\frac{1}{4}$; $£763\frac{3}{4}$]

7. A wholesaler of bicycles has a stock of 400 bicycles at his Manchester warehouse and a stock of 500 at his London warehouse. He receives an order for 400 from Birmingham and for 300 from Bristol. If he supplies Birmingham from Manchester his profit is £2.50 per bicycle but if he supplies Birmingham from London, it is £4 per bicycle. On the other hand, if he supplies Bristol from Manchester or from London his profit is £2 and £3 respectively per bicycle.

How should the order be allocated for maximum profit and what is this?

8. In a factory two types A and B of machines are to be installed. The following summarizes the information relevant to the machines:

	Type A	Type B	Maximum available
No. of men per machine	2	1	200
Labour cost per machine per week	£12	£10	£1 500
Maintenance cost per week per machine	£0.40	£0.50	£70
No. of machines	x	y	

What is the maximum number of machines which may be installed? If the weekly profits per machine are £6 and £4 for A and B respectively, what is the maximum weekly profit which may be obtained and how many machines of each type should be installed?

8 Matrices

8.1 The Order of a Matrix

A matrix is a rectangular array of numbers or other elements. The following are examples of matrices:

$$\begin{pmatrix} 1 & 2 & 3 \\ 0 & -1 & 4 \end{pmatrix} \qquad \begin{pmatrix} 1 & 0 & 0 & 0 \\ 0 & 1 & 0 & 0 \\ 0 & 0 & 1 & 0 \\ 0 & 0 & 0 & 1 \end{pmatrix} \qquad \begin{pmatrix} 2 \\ 1 \\ 3 \end{pmatrix} \qquad \begin{pmatrix} -2 & 1 & 0 & 4 \end{pmatrix}$$

If a matrix has m rows and n columns, it is said to be an m by n (written $m \times n$) matrix or a matrix of dimensions or of *order* $m \times n$. The above examples are respectively, from left to right, 2×3, 4×4, 3×1 and 1×4 matrices.

An $n \times 1$ matrix is called a *column* matrix. A $1 \times n$ matrix is called a *row* matrix. The third and fourth examples are respectively column and row matrices. A matrix having the same number of rows and columns is called a *square* matrix. An $n \times n$ matrix is a square matrix of order n. The second matrix above is a square matrix of order 4.

Matrices are usually denoted by bold capital letters **A**, **B**, **C**, ... In general the element in the ith row and the jth column of a matrix will be denoted by a_{ij}, so that the general form of, for example, a 3×4 matrix is

$$\begin{pmatrix} a_{11} & a_{12} & a_{13} & a_{14} \\ a_{21} & a_{22} & a_{23} & a_{24} \\ a_{31} & a_{32} & a_{33} & a_{34} \end{pmatrix}$$

Matrices arise quite naturally in mathematics. The following are some examples of their occurrence.

(1) *As matrices relating to a system of linear equations.*
The *coefficient* and *augmented* matrices of the system of equations

$$3x - 2y + z = 3$$
$$x + y - z = 1$$
$$2x + 3z = 0$$

are respectively

$$\begin{pmatrix} 3 & -2 & 1 \\ 1 & 1 & -1 \\ 2 & 0 & 3 \end{pmatrix} \quad \text{and} \quad \begin{pmatrix} 3 & -2 & 1 & 3 \\ 1 & 1 & -1 & 1 \\ 2 & 0 & 3 & 0 \end{pmatrix}$$

The order of the elements in these arrays is important because any change in this order would alter the system of equations for which these matrices stand. A matrix is thus an *ordered* rectangular array of elements.

(2) *As matrices relating to geometrical figures.*
An *incidence* matrix for a geometrical figure is formed by placing its rows opposite lines of the figure and its columns under points of the figure.
    The elements of the matrix are determined by the following rule. In the column under a particular point, enter 1 against any line in which that point lies, otherwise enter 0. For example, in the quadrilateral ABCD (Fig. 8.1), A lies on the lines AB, DA and AC but does not lie on BC or CD. The appropriate elements for column A are therefore as shown in the first column of the incidence matrix beside the figure, and the elements for columns B, C and D are determined similarly.

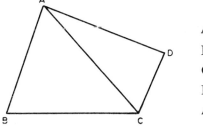

$$\begin{array}{c} \\ AB \\ BC \\ CD \\ DA \\ AC \end{array} \begin{array}{cccc} A & B & C & D \\ \begin{pmatrix} 1 & 1 & 0 & 0 \\ 0 & 1 & 1 & 0 \\ 0 & 0 & 1 & 1 \\ 1 & 0 & 0 & 1 \\ 1 & 0 & 1 & 0 \end{pmatrix} \end{array}$$

FIGURE 8.1

(3) *In the systematic tabulation of information.*

	I	II	III	IV
Arts	104	79	151	105
Science	152	203	122	156
Engineering	56	40	53	81

Matrix A (3 × 4)

	F	T	C	A
I	30	5	5	12
II	20	10	6	10
III	20	5	2	9
IV	10	5	2	0

Matrix B (4 × 4)

For example, matrix A shows the number of first, second, third and fourth year students at a university who were studying arts, science and engineering. Matrix B shows the numbers of attendances per university session for football, tennis, cricket and athletics by each student in the various years of study.

8.2 The Multiplication of Matrices

The systematic tabulations of information provided by matrices A and B above enable the answers to various questions to be obtained easily.

(*a*) How many games of football do arts students play per session?

$$104 \times 30 + 79 \times 20 + 151 \times 20 + 105 \times 10 = 8\,770$$

NOTE: To obtain this answer each of the elements of the first *row* of matrix A is multiplied by its corresponding element in the first column of matrix B and the products are added.

(*b*) How many attendances per session at athletics meetings are made by science students?

$$152 \times 12 + 203 \times 10 + 122 \times 9 + 156 \times 0 = 4\,952$$

This is obtained by multiplying the elements of the second *row* of A by those of the fourth *column* of B.

If each of the three rows of A is multiplied by each of the four columns of B in turn, the answers to all the 12 possible questions about the sporting activities of the university students are obtained.

These answers may be tabulated in the form of a 3 × 4 matrix **C**. This matrix is formed as follows: the answer obtained by multiplying row 1 of **A** by column 1 of **B** and adding the results, is placed in row 1 and column 1 of **C**; the answer obtained by multiplying row 2 of **A** by column 4 of **B** and adding the results, is placed in row 2 and column 4 of **C**; and so on.

$$\mathbf{C} = \begin{pmatrix} 8\,770 & 2\,741 & 1\,506 & 3\,397 \\ 12\,620 & 4\,302 & 2\,530 & 4\,952 \\ 3\,820 & 1\,403 & 788 & 1\,549 \end{pmatrix}$$

The matrix **C** is said to be the product of the matrix **A** by the matrix **B**, i.e. **C** = **A** × **B**, and **C** is seen to arise naturally as a systematic tabulation of the information about the sporting activities of the students.

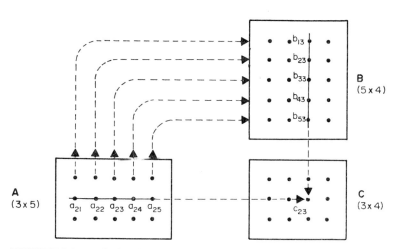

FIGURE 8.2 *Rule for the multiplication of two matrices.*

$$\begin{matrix} \mathbf{C} & = & \mathbf{A} & \times & \mathbf{B} \\ (3 \times 4) & & (3 \times 5) & & (5 \times 4) \end{matrix}$$

The schematic diagram (Fig. 8.2) illustrates the rule by which the elements of the product matrix **C** of two matrices **A** and **B** are found.

Any element of the product is found by multiplying the elements in the corresponding *row* of the first matrix by the elements in the corresponding *column* of the second matrix, element by element, and adding the results. For example, the element c_{23}, the element in the *second row* and the *third* column of **C**, is obtained by the multiplication of the second row of **A** by the third column of **B** and adding the results (Fig. 8.2):

$$c_{23} = a_{21}b_{13} + a_{22}b_{23} + a_{23}b_{33} + a_{24}b_{43} + a_{25}b_{53}$$

$$= \sum_{k=1}^{5} a_{2k}b_{k3}$$

In general,

$$c_{ij} = \sum_{k=1}^{5} a_{ik}b_{kj}$$

8.3   *Compatibility*

Since elements of the product **A** × **B** are formed by multiplying along the *rows* of **A** and down the *columns* of **B**, it follows that the number of *columns* of **A** must equal the number of *rows* of **B**.

Matrices which satisfy this criterion are said to be *compatible*, e.g. the matrices **A** and **B** (Fig. 8.2) have respectively 5 columns and 5 rows and are compatible for multiplication.

In general, a matrix **A** ($m \times n$) is compatible with matrix **B** ($n \times p$).

Matrices which are not compatible have no product or it may be said that their product is not defined.

EXAMPLE 8.1
(A) The product of two square (2 × 2) matrices:

$$\begin{pmatrix} 2 & 5 \\ 3 & 2 \end{pmatrix}\begin{pmatrix} 1 & 0 \\ 2 & -1 \end{pmatrix} = \begin{pmatrix} 12 & -5 \\ 7 & -2 \end{pmatrix}$$
$$\quad\; \mathbf{A} \qquad\quad \mathbf{B} \qquad\qquad \mathbf{AB}$$

$$\begin{pmatrix} 1 & 0 \\ 2 & -1 \end{pmatrix}\begin{pmatrix} 2 & 5 \\ 3 & 2 \end{pmatrix} = \begin{pmatrix} 2 & 5 \\ 1 & 8 \end{pmatrix}$$
$$\quad\; \mathbf{B} \qquad\quad \mathbf{A} \qquad\quad \mathbf{BA}$$

Note that **AB** ≠ **BA**.

(B)

$$\begin{pmatrix} 1 & 0 & 2 \\ 3 & 1 & 0 \end{pmatrix} \begin{pmatrix} -1 & 4 \\ 3 & -5 \\ 0 & 2 \end{pmatrix} = \begin{pmatrix} -1 & 8 \\ 0 & 7 \end{pmatrix}$$

A	**B**	**AB**
(2×3)	(3×2)	(2×2)

$$\begin{pmatrix} -1 & 4 \\ 3 & -5 \\ 0 & 2 \end{pmatrix} \begin{pmatrix} 1 & 0 & 2 \\ 3 & 1 & 0 \end{pmatrix} = \begin{pmatrix} 11 & 4 & -2 \\ -12 & -5 & 6 \\ 6 & 2 & 0 \end{pmatrix}$$

B	**A**	**BA**
(3×2)	(2×3)	(3×3)

Again $\mathbf{AB} \neq \mathbf{BA}$ and the products are of different orders.

In general, the product of two matrices $\mathbf{A}$ and $\mathbf{B}$ depends upon the order of multiplication so that, in the algebra of matrices, the commutative law for products does not hold.

The product of an $(m \times n)$ matrix by an $(n \times p)$ matrix is a matrix of order $(m \times p)$.

An $(m \times n)$ matrix $\mathbf{A}$ is compatible with an $(n \times m)$ matrix $\mathbf{B}$ for multiplication in either order but the product matrices will, in general, be different.

The product $\mathbf{AB}$ is an $(m \times m)$ matrix and the product $\mathbf{BA}$ is an $(n \times n)$ matrix.

8.4   Unit Matrix; Zero or Null Matrix

Clearly $\begin{pmatrix} 1 & 0 \\ 0 & 1 \end{pmatrix} \begin{pmatrix} a & b \\ c & d \end{pmatrix} = \begin{pmatrix} a & b \\ c & d \end{pmatrix} = \begin{pmatrix} a & b \\ c & d \end{pmatrix} \begin{pmatrix} 1 & 0 \\ 0 & 1 \end{pmatrix}$

Thus if any (2×2) matrix is premultiplied or postmultiplied by the matrix $\begin{pmatrix} 1 & 0 \\ 0 & 1 \end{pmatrix}$, it is unchanged.

The matrix $\begin{pmatrix} 1 & 0 \\ 0 & 1 \end{pmatrix}$ thus plays the role of unity in ordinary algebra, and is called the *unit matrix* of order 2 and is usually denoted by $\mathbf{I}$.

If we write $\mathbf{A} = \begin{pmatrix} a & b \\ c & d \end{pmatrix}$, then $\mathbf{IA} = \mathbf{A} = \mathbf{AI}$.

The unit matrix of order 3 is $\begin{pmatrix} 1 & 0 & 0 \\ 0 & 1 & 0 \\ 0 & 0 & 1 \end{pmatrix}$. All its elements are zero except

those in the leading diagonal of the array which are each unity. Unit matrices of higher orders have a similar pattern. Unit matrices of any order will be denoted by $\mathbf{I}$, the order being specified.

A *zero or null matrix* has all its elements zero and is denoted by $\mathbf{0}$ so that

$$\mathbf{0A} = \mathbf{A0} = \mathbf{0}$$

Note that
(1) $\mathbf{AB} = \mathbf{0}$ does not necessarily imply that either $\mathbf{A} = \mathbf{0}$ or $\mathbf{B} = \mathbf{0}$ as in ordinary algebra, e.g.

$$\text{if } \mathbf{A} = \begin{pmatrix} -1 & 1 \\ 0 & 0 \end{pmatrix} \quad \text{and} \quad \mathbf{B} = \begin{pmatrix} 0 & 1 \\ 0 & 1 \end{pmatrix} \quad \text{then } \mathbf{AB} = \mathbf{0}$$

(2) $\mathbf{AB} = \mathbf{AC}$ does not necessarily imply that either $\mathbf{A} = \mathbf{0}$ or $\mathbf{B} = \mathbf{C}$, e.g.

$$\text{if } \mathbf{A} = \begin{pmatrix} 2 & 0 \\ 1 & 0 \end{pmatrix} \quad \mathbf{B} = \begin{pmatrix} 1 & 2 \\ 0 & 0 \end{pmatrix} \quad \text{and} \quad \mathbf{C} = \begin{pmatrix} 1 & 2 \\ 1 & 0 \end{pmatrix}$$

$$\mathbf{AB} = \begin{pmatrix} 2 & 0 \\ 1 & 0 \end{pmatrix}\begin{pmatrix} 1 & 2 \\ 0 & 0 \end{pmatrix} = \begin{pmatrix} 2 & 4 \\ 1 & 2 \end{pmatrix}$$

$$\mathbf{AC} = \begin{pmatrix} 2 & 0 \\ 1 & 0 \end{pmatrix}\begin{pmatrix} 1 & 2 \\ 1 & 0 \end{pmatrix} = \begin{pmatrix} 2 & 4 \\ 1 & 2 \end{pmatrix}$$

i.e. $\mathbf{AB} = \mathbf{BC}$ but $\mathbf{A} \neq \mathbf{0}$ and $\mathbf{B} \neq \mathbf{C}$.

8.5 Equality, Addition, Multiplication by a Scalar

Two matrices are equal if and only if they are of the same order and their corresponding elements are equal:

$$\begin{pmatrix} a_1 & b_1 \\ c_1 & d_1 \end{pmatrix} = \begin{pmatrix} a_2 & b_2 \\ c_2 & d_2 \end{pmatrix} \Leftrightarrow \begin{matrix} a_1 = a_2, & b_1 = b_2 \\ c_1 = c_2, & d_1 = d_2 \end{matrix}$$

The sum of two matrices of the same order is defined to be a matrix whose elements are the sums of the corresponding elements of these two matrices:

$$\begin{pmatrix} a_{11} & a_{12} & a_{13} \\ a_{21} & a_{22} & a_{23} \end{pmatrix} + \begin{pmatrix} b_{11} & b_{12} & b_{13} \\ b_{21} & b_{22} & b_{23} \end{pmatrix} = \begin{pmatrix} a_{11} + b_{11} & a_{12} + b_{12} & a_{13} + b_{13} \\ a_{21} + b_{21} & a_{22} + b_{22} & a_{23} + b_{23} \end{pmatrix}$$

The product of a matrix by a number (scalar) k is defined to be the matrix derived from the first by multiplying each of its elements by k:

$$\mathbf{A} = \begin{pmatrix} a & b \\ c & d \end{pmatrix}$$

$$k\mathbf{A} = \begin{pmatrix} ka & kb \\ kc & kd \end{pmatrix}$$

EXERCISES 8.1

1. Perform the following additions (where possible)

(a) $(2 \quad 1 \quad -1) + (0 \quad 4 \quad 3)$

(b) $\begin{pmatrix} 1 \\ 0 \\ 2 \end{pmatrix} + (5 \quad -2 \quad 4)$

(c) $\begin{pmatrix} 2 \\ -1 \\ 4 \end{pmatrix} + \begin{pmatrix} 0 \\ -2 \\ -3 \end{pmatrix}$

(d) $\begin{pmatrix} 2 & 2 & 5 \\ 1 & 0 & -2 \\ -3 & 4 & -3 \end{pmatrix} + \begin{pmatrix} 0 & 1 & 3 \\ -2 & 4 & -1 \\ 2 & -3 & 3 \end{pmatrix}$

(e) $a(1 \quad -1 \quad 1) + b(-2 \quad 0 \quad -1) + c(2 \quad -1 \quad 0)$

2. Find $\mathbf{A} - \mathbf{B}$, where $\mathbf{A} - \mathbf{B}$ is defined to be $\mathbf{A} + (-\mathbf{B})$

(a) $\mathbf{A} = \begin{pmatrix} 3 \\ 1 \\ 2 \end{pmatrix}$, $\mathbf{B} = \begin{pmatrix} 2 \\ -1 \\ 2 \end{pmatrix}$ (b) $\mathbf{A} = -\begin{pmatrix} 1 & 2 \\ 3 & 4 \end{pmatrix}$, $\mathbf{B} = \begin{pmatrix} 3 & 1 \\ 2 & 4 \end{pmatrix}$

(c) $\mathbf{A} = \begin{pmatrix} x & y \\ 0 & 1 \\ y & 2 \end{pmatrix}$, $\mathbf{B} = \begin{pmatrix} -1 & 7 \\ x & 1 \\ 2x & -2 \end{pmatrix}$

3. If $\mathbf{A} = \begin{pmatrix} 2 & -3 \\ 1 & 4 \end{pmatrix}$ and $\mathbf{B} = \begin{pmatrix} 3 & -1 \\ 2 & -5 \end{pmatrix}$

   find the matrices $\mathbf{A} + \mathbf{B}$, $\mathbf{B} - \mathbf{A}$, $3\mathbf{A} + 2\mathbf{B}$, $3\mathbf{B} - 2\mathbf{A}$.

4. If $\mathbf{A} = \begin{pmatrix} a & -2b & c \\ 2a & b & -3c \end{pmatrix}$, $\mathbf{B} = \begin{pmatrix} -a & b & -c \\ 4a & -3b & 2c \end{pmatrix}$ and $\mathbf{C} = \begin{pmatrix} 3a & -b & 2c \\ a & b & -c \end{pmatrix}$

   find the matrix $2\mathbf{A} - 3\mathbf{B} + \mathbf{C}$.

5. An electrical network consists of nodes a, b, c, d, and branches 1, 2, 3, 4, 5, 6 (Fig. 8.3). Construct an incidence matrix in which the rows are labelled a, b, c, d and the columns 1, 2, 3, 4, 5, 6. The node a lies on the branches 1, 3, 5 and no others. In line a therefore 1 is written in columns 1, 3, 5 and 0 in the others. Complete the matrix.

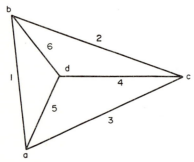

$$\text{Matrix} = \dfrac{\begin{array}{c|cccccc} & 1 & 2 & 3 & 4 & 5 & 6 \\ \hline a & 1 & 0 & 1 & 0 & 1 & 0 \\ b & & & & & & \\ c & & & & & & \\ d & & & & & & \end{array}}{}$$

FIGURE 8.3

6. If Fig. 8.3 is taken to represent the roads which connect four towns, a, b, c, d. Show that the matrix

$$\mathbf{A} = \begin{array}{c} \\ a \\ b \\ c \\ d \end{array}\begin{array}{c} \begin{array}{cccc} a & b & c & d \end{array} \\ \begin{pmatrix} 0 & 1 & 1 & 1 \\ 1 & 0 & 1 & 1 \\ 1 & 1 & 0 & 1 \\ 1 & 1 & 1 & 0 \end{pmatrix} \end{array}$$

tabulates the direct routes between towns, e.g. 1 entered in row c and in column d shows that there is a direct route from c to d; 0 indicates there is no direct route from a to a, etc. Show that the matrix $\mathbf{B}$ which tabulates the number of routes between two towns which pass through one other town (and not more) is

$$\mathbf{B} = \begin{array}{c} \\ a \\ b \\ c \\ d \end{array}\begin{array}{c} \begin{array}{cccc} a & b & c & d \end{array} \\ \begin{pmatrix} 3 & 2 & 2 & 2 \\ 2 & 3 & 2 & 2 \\ 2 & 2 & 3 & 2 \\ 2 & 2 & 2 & 3 \end{pmatrix} \end{array}$$

The entry 3 in column 1 and row 1 indicates that there are 3 routes from a to a, namely a to b, a to c, and a to d and return by the same road. Show that $B = AA = A^2$ and explain this.

7. A firm, manufacturing refrigerators, washing machines, spin-driers and mixers, has four branches (B_1, B_2, B_3, B_4) in each of the towns London, Birmingham, Manchester and Glasgow. Matrix 1 gives the numbers of salesmen working in each of the branches in these four towns.

   Matrix 2 shows the numbers of refrigerators (R), washing machines (W), spin-driers (S) and mixers (M) sold per day by each of the salesmen in each of the branches in the four towns.

   Express in matrix form the total daily sales of each of the products in each of the four towns.

	B_1	B_2	B_3	B_4
London	3	2	2	3
Birmingham	4	3	2	2
Manchester	2	2	1	3
Glasgow	1	3	3	2

Matrix 1

	R	W	S	M
B_1	5	4	5	3
B_2	6	5	5	4
B_3	4	3	4	2
B_4	5	6	6	3

Matrix 2

8. Multiply the following 2×2 matrices in both orders.

(a) $\begin{pmatrix} 1 & 2 \\ 3 & 4 \end{pmatrix}\begin{pmatrix} 5 & 2 \\ 1 & 3 \end{pmatrix}$ (b) $\begin{pmatrix} 5 & 3 \\ 2 & 4 \end{pmatrix}\begin{pmatrix} 6 & -2 \\ 1 & 2 \end{pmatrix}$

(c) $\begin{pmatrix} 4 & -3 \\ -2 & 5 \end{pmatrix}\begin{pmatrix} 5 & 3 \\ 2 & 4 \end{pmatrix}$ (d) $\begin{pmatrix} 5 & 2 \\ 10 & 4 \end{pmatrix}\begin{pmatrix} 4 & -2 \\ -10 & 5 \end{pmatrix}$

(e) $\begin{pmatrix} 3 & -2 \\ -7 & 5 \end{pmatrix}\begin{pmatrix} 5 & 2 \\ 7 & 3 \end{pmatrix}$

Show that the products are not commutative except in (d) and (e). Note that (d) illustrates that $AB = 0 \not\Rightarrow A = 0$ or $B = 0$.

$\left[(a)\ \begin{pmatrix} 7 & 8 \\ 19 & 18 \end{pmatrix}\quad (b)\ \begin{pmatrix} 33 & -4 \\ 16 & 4 \end{pmatrix}\quad (c)\ \begin{pmatrix} 14 & 0 \\ 0 & 14 \end{pmatrix}\quad (d)\ \begin{pmatrix} 0 & 0 \\ 0 & 0 \end{pmatrix}\quad (e)\ \begin{pmatrix} 1 & 0 \\ 0 & 1 \end{pmatrix} \right]$

9. Multiply in both orders, if possible.

(a) $\begin{pmatrix} 4 & 2 \\ 1 & -3 \\ 5 & 7 \end{pmatrix}\begin{pmatrix} -2 & 5 \\ 8 & -1 \end{pmatrix}$ (b) $\begin{pmatrix} 2 & 0 & -1 \\ 3 & -1 & 2 \end{pmatrix}\begin{pmatrix} 0 & 1 & -1 \\ 3 & 0 & -2 \\ 2 & -3 & 1 \end{pmatrix}$

(c) $\begin{pmatrix} 0 & 0 & 3 \\ 0 & 4 & 2 \\ 1 & 3 & 5 \end{pmatrix}\begin{pmatrix} 1 & 3 & 0 \\ 2 & 0 & 0 \\ 0 & 0 & 0 \end{pmatrix}$ (d) $(1 \quad -2 \quad 4)\begin{pmatrix} -2 & 4 \\ 3 & 5 \\ -6 & 1 \end{pmatrix}$

(e) $\begin{pmatrix} -2 & 3 & -6 \\ 4 & 5 & 1 \end{pmatrix} \begin{pmatrix} 1 \\ -2 \\ 4 \end{pmatrix}$ (f) $\begin{pmatrix} 3 & 5 & 1 \\ -1 & 2 & 0 \end{pmatrix} \begin{pmatrix} 2/11 & -5/11 \\ 1/11 & 3/11 \\ 0 & 0 \end{pmatrix}$

$\left[\text{(a)} \begin{pmatrix} 8 & 18 \\ -26 & 8 \\ 46 & 18 \end{pmatrix} \quad \text{(b)} \begin{pmatrix} -2 & 5 & -3 \\ 1 & -3 & 1 \end{pmatrix} \quad \text{(c)} \begin{pmatrix} 0 & 0 & 0 \\ 8 & 0 & 0 \\ 7 & 3 & 0 \end{pmatrix}; \begin{pmatrix} 0 & 12 & 9 \\ 0 & 0 & 6 \\ 0 & 0 & 0 \end{pmatrix} \right.$

(d) $(-32 \quad -2)$ (e) $\begin{pmatrix} -32 \\ -2 \end{pmatrix}$ (f) $\begin{pmatrix} 1 & 0 \\ 0 & 1 \end{pmatrix}; \begin{pmatrix} 1 & 0 & 2/11 \\ 0 & 1 & 1/11 \\ 0 & 0 & 0 \end{pmatrix} \right]$

10. Find **AB** and **BA** if

(a) $\mathbf{A} = (-4 \quad -1 \quad 2)$ and $\mathbf{B} = \begin{pmatrix} 2 \\ 1 \\ -1 \end{pmatrix}$

(b) $\mathbf{A} = \begin{pmatrix} -3 & 0 & 1 \\ 4 & 2 & -1 \\ 2 & -2 & 0 \end{pmatrix}$ and $\mathbf{B} = \begin{pmatrix} 0 & 1 & 2 \\ 5 & 0 & -3 \\ -2 & 1 & 4 \end{pmatrix}$

Verify that $(\mathbf{AB})\mathbf{A} = \mathbf{A}(\mathbf{BA})$.

$\left[\text{(a)} \ (-11), \begin{pmatrix} -8 & -2 & 4 \\ -4 & -1 & 2 \\ 4 & 1 & -2 \end{pmatrix}; \quad (44 \quad 11 \quad -22) \right.$

(b) $\mathbf{AB} = \begin{pmatrix} -2 & -2 & -2 \\ 12 & 3 & -2 \\ -10 & 2 & 10 \end{pmatrix}; \mathbf{BA} = \begin{pmatrix} 8 & -2 & -1 \\ -21 & 6 & 5 \\ 18 & -6 & -3 \end{pmatrix}$

$(\mathbf{AB})\mathbf{A} = \mathbf{A}(\mathbf{BA}) = \begin{pmatrix} -6 & 0 & 0 \\ -28 & 10 & 9 \\ 58 & -16 & -12 \end{pmatrix} \right]$

11. Show that

(a) $\begin{pmatrix} 2 & -3 \\ 5 & 1 \end{pmatrix} \begin{pmatrix} x \\ y \end{pmatrix} = \begin{pmatrix} 2x - 3y \\ 5x + y \end{pmatrix}$

(b) $\begin{pmatrix} 2 & -1 & 0 \\ 1 & 5 & -2 \\ 3 & 0 & 1 \end{pmatrix} \begin{pmatrix} x \\ y \\ z \end{pmatrix} = \begin{pmatrix} 2x - y \\ x + 5y - 2z \\ 3x + z \end{pmatrix}$

(c) $(x \quad y)\begin{pmatrix} a & h \\ h & b \end{pmatrix}\begin{pmatrix} x \\ y \end{pmatrix} = ax^2 + 2hxy + by^2$

(d) $(x \quad y \quad z)\begin{pmatrix} a & h & g \\ h & b & f \\ g & f & c \end{pmatrix}\begin{pmatrix} x \\ y \\ z \end{pmatrix} = ax^2 + by^2 + cz^2 + 2fyz + 2gzx + 2hxy$

8.6 The Determinant of a Square Matrix

The *determinant* of an $n \times n$ matrix of real numbers

$$A = \begin{pmatrix} a_{11} & a_{12} & \cdots & a_{1n} \\ a_{21} & a_{22} & \cdots & a_{2n} \\ \cdot & \cdot & \cdot & \cdot \\ \cdot & \cdot & \cdot & \cdot \\ \cdot & \cdot & \cdot & \cdot \\ a_{n1} & a_{n2} & \cdot & a_{nn} \end{pmatrix}$$

is denoted by

$$|A| = \begin{vmatrix} a_{11} & a_{12} & \cdots & a_{1n} \\ a_{21} & a_{22} & \cdots & a_{2n} \\ \cdot & \cdot & \cdot & \cdot \\ \cdot & \cdot & \cdot & \cdot \\ \cdot & \cdot & \cdot & \cdot \\ a_{n1} & a_{n2} & \cdot & a_{nn} \end{vmatrix}$$

and is said to be a determinant of *order n*.

A rule will now be stated by which a value may be assigned to a determinant. We may, without ambiguity, denote the value of the determinant $|A|$ of the matrix A by $|A|$ also.

Let α_{ij} be the matrix obtained by deleting the ith row and the jth column of the matrix A above. α_{ij} is thus a square matrix of order $(n - 1)$.

The determinant $|\alpha_{ij}|$ is called the *minor* in the determinant $|A|$ corresponding to the element a_{ij}. We define the *cofactor* A_{ij} of the element a_{ij} to be $(-1)^{i+j}$ times the value of the minor $|\alpha_{ij}|$, i.e.

$$A_{ij} = (-1)^{i+j} |\alpha_{ij}|$$

The value assigned to the determinant $|\mathbf{A}|$ is given by

$$|\mathbf{A}| = \sum_{j=1}^{n} a_{ij} A_{ij}$$

being expressed in terms of the elements of the ith row and their cofactors.

It can be shown that the same value of $|\mathbf{A}|$ will be derived from the above formula irrespective of the particular row in terms of which it is evaluated. It is also well known that if the products obtained by multiplying each element of any row by the corresponding cofactor of another row and these products are added, the sum will be zero:

$$\sum_{r=1}^{n} a_{ik} A_{jk} = 0 \qquad i \neq j$$

For a matrix of order 1 comprising the single element a_{11}, the value of its determinant is a_{11}. For the 2 × 2 matrix $\begin{pmatrix} a_{11} & a_{12} \\ a_{21} & a_{22} \end{pmatrix}$, its determinant is

$$\begin{vmatrix} a_{11} & a_{12} \\ a_{21} & a_{22} \end{vmatrix} = a_{11} A_{11} + a_{12} A_{12}$$

$$= a_{11} a_{22} + a_{12}(-a_{21})$$

For example

$$\begin{vmatrix} a & b \\ c & d \end{vmatrix} = ad - bc$$

$$\begin{vmatrix} 3 & -1 \\ -2 & 4 \end{vmatrix} = 3 \times 4 - (-1) \times (-2) = 12 - 2 = 10$$

For a (3 × 3) matrix

$$\begin{vmatrix} a_{11} & a_{12} & a_{13} \\ a_{21} & a_{22} & a_{23} \\ a_{31} & a_{32} & a_{33} \end{vmatrix} = a_{11} A_{11} + a_{12} A_{12} + a_{13} A_{13}$$

$$= a_{11} \begin{vmatrix} a_{22} & a_{23} \\ a_{32} & a_{33} \end{vmatrix} - a_{12} \begin{vmatrix} a_{21} & a_{23} \\ a_{31} & a_{33} \end{vmatrix} + a_{13} \begin{vmatrix} a_{21} & a_{22} \\ a_{31} & a_{32} \end{vmatrix}$$

$$= a_{11}(a_{22}a_{33} - a_{23}a_{32}) - a_{12}(a_{21}a_{33} - a_{23}a_{31})$$

$$+ a_{13}(a_{21}a_{32} - a_{22}a_{31})$$

For example

$$\begin{vmatrix} -2 & -5 & 1 \\ 3 & -2 & 2 \\ 1 & 4 & -6 \end{vmatrix} = -2\begin{vmatrix} -2 & 2 \\ 4 & -6 \end{vmatrix} - (-5)\begin{vmatrix} 3 & 2 \\ 1 & -6 \end{vmatrix} + (1)\begin{vmatrix} 3 & -2 \\ 1 & 4 \end{vmatrix}$$

$$= -2(12 - 8) + 5(-18 - 2) + (12 + 2)$$

$$= -94$$

The determinant of a 4 × 4 matrix may be expressed in terms of 4 determinants of order 3 and so on.

A matrix, whose determinant is zero, is said to be a *singular* matrix.

One of the rules for multiplying two determinants of the same order is precisely the rule (stated in Section 8.2) for multiplying two square matrices of the same order. Hence

$$|AB| = |A|\,|B|$$

where **A**, **B** are square matrices of the same order.

Also $|AB| = |BA|$ even though, in general, $AB \neq BA$.

EXERCISES 8.2

1. Find the values of the determinants of the following matrices:—

(a) $\begin{pmatrix} 2 & 3 \\ 1 & 4 \end{pmatrix}$ (b) $\begin{pmatrix} -2 & 4 \\ 5 & -7 \end{pmatrix}$ (c) $\begin{pmatrix} 3 & 2 \\ 6 & 4 \end{pmatrix}$

(d) $\begin{pmatrix} 1 & 0 & 0 \\ 0 & 1 & 0 \\ 0 & 0 & 1 \end{pmatrix}$ (e) $\begin{pmatrix} 2 & 0 & 1 \\ 3 & 4 & 0 \\ 0 & 2 & 1 \end{pmatrix}$ (f) $\begin{pmatrix} -4 & 1 & -2 \\ 3 & -5 & 7 \\ -2 & 4 & -5 \end{pmatrix}$

[5, −6, 0, 1, 14, 9]

2. Show that $\begin{pmatrix} a & h & g \\ h & b & f \\ g & f & c \end{pmatrix}$ is a singular matrix if $a = b = h$ and $f = g$.

3. Verify that $|AB| = |A|\,|B|$ if

(i) $A = \begin{pmatrix} -2 & 3 \\ 4 & -5 \end{pmatrix}$ and $B = \begin{pmatrix} 1 & -4 \\ -3 & 2 \end{pmatrix}$

[$|A| = -2$; $|B| = -10$; $|AB| = 20$]

(ii) $\mathbf{A} = \begin{pmatrix} 4 & 0 & 1 \\ -5 & 2 & 4 \\ 3 & -1 & 6 \end{pmatrix}$ and $\mathbf{B} = \begin{pmatrix} 1 & 4 & -2 \\ 0 & 3 & 6 \\ -5 & 2 & -1 \end{pmatrix}$

$$\left[|\mathbf{A}| = 63; |\mathbf{B}| = -165; |\mathbf{A}||\mathbf{B}| = -10\,395; \right.$$

$$\left. \mathbf{AB} = \begin{pmatrix} -1 & 18 & -9 \\ -25 & -6 & 18 \\ -27 & 21 & -18 \end{pmatrix}; |\mathbf{AB}| = -10\,395 \right]$$

8.7   Inverse Matrix

It is easily verified that

$$\begin{pmatrix} 3 & 1 \\ 5 & 2 \end{pmatrix} \begin{pmatrix} 2 & -1 \\ -5 & 3 \end{pmatrix} = \begin{pmatrix} 1 & 0 \\ 0 & 1 \end{pmatrix} = \mathbf{I} \quad \text{(unit matrix)}$$

$$\begin{pmatrix} 2 & -1 \\ -5 & 3 \end{pmatrix} \begin{pmatrix} 3 & 1 \\ 5 & 2 \end{pmatrix} = \begin{pmatrix} 1 & 0 \\ 0 & 1 \end{pmatrix} = \mathbf{I}$$

When multiplied together, in either order, these two matrices yield a unit matrix.

Each matrix is said to be the *inverse* of the other. The inverse of a matrix $\mathbf{A}$ will be denoted by $\mathbf{A}^{-1}$, and is such that

$$\mathbf{A}^{-1}\mathbf{A} = \mathbf{A}\mathbf{A}^{-1} = \mathbf{I}$$

THE INVERSE OF A 2 × 2 MATRIX

Let the inverse of matrix $\mathbf{A} = \begin{pmatrix} a & b \\ c & d \end{pmatrix}$ be $\begin{pmatrix} x & y \\ p & q \end{pmatrix}$. Then

$$\begin{pmatrix} x & y \\ p & q \end{pmatrix} \begin{pmatrix} a & b \\ c & d \end{pmatrix} = \begin{pmatrix} 1 & 0 \\ 0 & 1 \end{pmatrix}$$

$$\begin{pmatrix} ax + cy & bx + dy \\ pa + qc & pb + qd \end{pmatrix} = \begin{pmatrix} 1 & 0 \\ 0 & 1 \end{pmatrix}$$

$$ax + cy = 1 \qquad bx + dy = 0$$

$$pa + qc = 0 \qquad pb + qd = 1$$

Therefore

$$x = -\frac{d}{b}y \qquad y\left(-\frac{ad}{b} + c\right) = 1$$

Hence

$$y = \frac{-b}{ad - bc} \quad \text{and} \quad x = \frac{d}{ad - bc}$$

provided $ad - bc \neq 0$.
  Similarly

$$p = \frac{c}{ad - bc} \qquad q = \frac{a}{ad - bc}$$

Therefore

$$\mathbf{A}^{-1} = \frac{1}{(ad - bc)}\begin{pmatrix} d & -b \\ -c & a \end{pmatrix} = \frac{1}{|\mathbf{A}|}\begin{pmatrix} d & -b \\ -c & a \end{pmatrix}$$

where $|\mathbf{A}| = ad - bc = \begin{vmatrix} a & b \\ c & d \end{vmatrix}$ provided $|\mathbf{A}| \neq 0$, i.e. provided $\mathbf{A}$ is not a singular matrix. It is easily verified that

$$\frac{1}{|\mathbf{A}|}\begin{pmatrix} d & -b \\ -c & a \end{pmatrix}\begin{pmatrix} a & b \\ c & d \end{pmatrix} = \frac{1}{|\mathbf{A}|}\begin{pmatrix} |\mathbf{A}| & 0 \\ 0 & |\mathbf{A}| \end{pmatrix} = \begin{pmatrix} 1 & 0 \\ 0 & 1 \end{pmatrix} = \mathbf{I}$$

and that

$$\frac{1}{|\mathbf{A}|}\begin{pmatrix} a & b \\ c & d \end{pmatrix}\begin{pmatrix} d & -b \\ -c & a \end{pmatrix} = \frac{1}{|\mathbf{A}|}\begin{pmatrix} |\mathbf{A}| & 0 \\ 0 & |\mathbf{A}| \end{pmatrix} = \begin{pmatrix} 1 & 0 \\ 0 & 1 \end{pmatrix} = \mathbf{I}$$

i.e.

$$\mathbf{A}^{-1}\mathbf{A} = \mathbf{A}\mathbf{A}^{-1} = \mathbf{I}$$

Rule: The inverse of a 2 × 2 matrix is obtained by interchanging the top left and the bottom right elements, changing the signs of the other two elements and multiplying by the inverse of the determinant of the matrix.

EXAMPLE 8.2
Find the inverses of the following matrices:

(a) $A = \begin{pmatrix} 5 & 4 \\ 2 & 3 \end{pmatrix}$

$$|A| = \begin{vmatrix} 5 & 4 \\ 2 & 3 \end{vmatrix} = 5 \times 3 - 4 \times 2 = 7$$

$$A^{-1} = \frac{1}{7} \begin{pmatrix} 3 & -4 \\ -2 & 5 \end{pmatrix}$$

(b) $B = \begin{pmatrix} 3 & -2 \\ -4 & 1 \end{pmatrix}$

$$|B| = \begin{vmatrix} 3 & -2 \\ -4 & 1 \end{vmatrix} = 3 \times 1 - (-4) \times (-2) = -5$$

$$B^{-1} = -\frac{1}{5} \begin{pmatrix} 1 & 2 \\ 4 & 3 \end{pmatrix}$$

(c) $C = \begin{pmatrix} 4 & 2 \\ 6 & 3 \end{pmatrix}$

$$|C| = \begin{vmatrix} 4 & 2 \\ 6 & 3 \end{vmatrix} = 4 \times 3 - 6 \times 2 = 0$$

C has no inverse; it is a singular matrix.

EXERCISES 8.3

1. Find, where possible, the inverses of

(a) $\begin{pmatrix} 3 & 4 \\ 5 & 7 \end{pmatrix}$ (b) $\begin{pmatrix} 3 & -1 \\ -9 & 3 \end{pmatrix}$

(c) $\begin{pmatrix} 2 & 3 \\ 1 & 5 \end{pmatrix}$ (d) $\begin{pmatrix} 4 & -1 \\ 1 & 2 \end{pmatrix}$

2. Let the letters of the alphabet represent numbers as follows:

A	B	C	D	E	F	G	H	I	J	K	L	M	N	O	P	Q
20	17	8	7	2	6	9	12	10	15	18	24	3	16	25	14	11

R	S	T	U	V	W	X	Y	Z
1	5	4	23	19	22	13	21	26

Using this code any group of four letters may be represented by a 2 × 2 matrix, e.g.

TERM becomes $\begin{pmatrix} 4 & 2 \\ 1 & 3 \end{pmatrix}$

This code would probably not be difficult to "break". To make it more difficult to break, we choose any encoder, e.g. $\begin{pmatrix} 5 & 2 \\ 2 & 1 \end{pmatrix}$, and use this matrix to premultiply the first matrix thus:

$$\begin{pmatrix} 5 & 2 \\ 2 & 1 \end{pmatrix}\begin{pmatrix} 4 & 2 \\ 1 & 3 \end{pmatrix} = \begin{pmatrix} 22 & 16 \\ 9 & 7 \end{pmatrix}$$

The message TERM is now sent off as 22, 16, 9, 7

How is this to be decoded at the receiving end? By premultiplying by the decoder $\begin{pmatrix} 1 & -2 \\ -2 & 5 \end{pmatrix}$ thus:

$$\begin{pmatrix} 1 & -2 \\ -2 & 5 \end{pmatrix}\begin{pmatrix} 22 & 16 \\ 9 & 7 \end{pmatrix} = \begin{pmatrix} 4 & 2 \\ 1 & 3 \end{pmatrix} \rightarrow \text{TERM}$$

(i) What is the relationship of the decoder $\begin{pmatrix} 1 & -2 \\ -2 & 5 \end{pmatrix}$ to the encoder $\begin{pmatrix} 5 & 2 \\ 2 & 1 \end{pmatrix}$?

(ii) Show that the same encoder and decoder will "work" with the messages FISH, TEST, RAIN.

(iii) Why will they "work" with any group of four letters?

8.8   Solution of Linear Simultaneous Equations

Let us consider the matrix solution of the simplest case, namely two equations in two unknowns x and y of the general form

$$ax + by = p$$
$$cx + dy = q$$

These equations may be written in matrix form thus:

$$\begin{pmatrix} ax + by \\ cx + dy \end{pmatrix} = \begin{pmatrix} p \\ q \end{pmatrix}$$

or

$$\begin{pmatrix} a & b \\ c & d \end{pmatrix}\begin{pmatrix} x \\ y \end{pmatrix} = \begin{pmatrix} p \\ q \end{pmatrix}$$

Before solving the general case, we solve some particular cases.

(1)
$$3x + y = 5$$
$$5x + 2y = 8$$

$$\begin{pmatrix} 3 & 1 \\ 5 & 2 \end{pmatrix}\begin{pmatrix} x \\ y \end{pmatrix} = \begin{pmatrix} 5 \\ 8 \end{pmatrix}$$

Premultiply each side of the last equation by the inverse of the coefficient matrix $\begin{pmatrix} 3 & 1 \\ 5 & 2 \end{pmatrix}$, i.e. by $\begin{pmatrix} 2 & -1 \\ -5 & 3 \end{pmatrix}$. Therefore

$$\begin{pmatrix} 2 & -1 \\ -5 & 3 \end{pmatrix}\begin{pmatrix} 3 & 1 \\ 5 & 2 \end{pmatrix}\begin{pmatrix} x \\ y \end{pmatrix} = \begin{pmatrix} 2 & -1 \\ -5 & 3 \end{pmatrix}\begin{pmatrix} 5 \\ 8 \end{pmatrix}$$

$$\begin{pmatrix} 1 & 0 \\ 0 & 1 \end{pmatrix}\begin{pmatrix} x \\ y \end{pmatrix} = \begin{pmatrix} 2 \\ -1 \end{pmatrix}$$

$$\begin{pmatrix} x \\ y \end{pmatrix} = \begin{pmatrix} 2 \\ -1 \end{pmatrix}$$

Therefore $x = 2, y = -1$.

(2)
$$3x + 4y = 1$$
$$5x - 7y = -12$$

$$\begin{pmatrix} 3 & 4 \\ 5 & -7 \end{pmatrix}\begin{pmatrix} x \\ y \end{pmatrix} = \begin{pmatrix} 1 \\ -12 \end{pmatrix}$$

$$-\frac{1}{41}\begin{pmatrix} -7 & -4 \\ -5 & 3 \end{pmatrix}\begin{pmatrix} 3 & 4 \\ 5 & -7 \end{pmatrix}\begin{pmatrix} x \\ y \end{pmatrix} = -\frac{1}{41}\begin{pmatrix} -7 & -4 \\ -5 & 3 \end{pmatrix}\begin{pmatrix} 1 \\ -12 \end{pmatrix}$$

$$\begin{pmatrix} x \\ y \end{pmatrix} = \frac{1}{41}\begin{pmatrix} 7 & 4 \\ 5 & -3 \end{pmatrix}\begin{pmatrix} 1 \\ -12 \end{pmatrix} = \frac{1}{41}\begin{pmatrix} -41 \\ 41 \end{pmatrix} = \begin{pmatrix} -1 \\ 1 \end{pmatrix}$$

Therefore $x = -1, y = 1$.

Continuing the solution of the general case, we premultiply each side of the last equation obtained by the inverse of the coefficient matrix $\mathbf{A} = \begin{pmatrix} a & b \\ c & d \end{pmatrix}$, i.e.

$$\text{by } \frac{1}{|\mathbf{A}|}\begin{pmatrix} d & -b \\ -c & a \end{pmatrix} \text{ provided } |\mathbf{A}| = \begin{vmatrix} a & b \\ c & d \end{vmatrix} \neq 0$$

Therefore

$$\frac{1}{|\mathbf{A}|}\begin{pmatrix} d & -b \\ -c & a \end{pmatrix}\begin{pmatrix} a & b \\ c & d \end{pmatrix}\begin{pmatrix} x \\ y \end{pmatrix} = \frac{1}{|\mathbf{A}|}\begin{pmatrix} d & -b \\ -c & a \end{pmatrix}\begin{pmatrix} p \\ q \end{pmatrix}$$

$$\begin{pmatrix} x \\ y \end{pmatrix} = \frac{1}{|\mathbf{A}|}\begin{pmatrix} pd - qb \\ aq - cp \end{pmatrix}$$

$$x = \frac{pd - qb}{|\mathbf{A}|} = \frac{\begin{vmatrix} p & b \\ q & d \end{vmatrix}}{\begin{vmatrix} a & b \\ c & d \end{vmatrix}} \qquad y = \frac{aq - cp}{|\mathbf{A}|} = \frac{\begin{vmatrix} a & p \\ c & q \end{vmatrix}}{\begin{vmatrix} a & b \\ c & d \end{vmatrix}}$$

provided $\begin{vmatrix} a & b \\ c & d \end{vmatrix} = ad - bc \neq 0$ (Cramer's Rule).

The solution fails when $ad - bc = 0$, i.e. when $a/b = c/d$. In solving a pair of equations graphically, the coordinates of the point of intersection of their straight line graphs are the solutions.

The slopes of the two straight lines $ax + by = p$ and $cx + dy = q$ are $-a/b$ and $-c/d$ respectively. It is clear that when $|\mathbf{A}| = 0$, the lines are parallel and no solution of the corresponding equations could be expected.

EXERCISES 8.4
Solve where possible the following pairs of simultaneous equations:

1. $3x - y = 5$
   $x + 4y = 6$
   $[2, 1]$

2. $5x + 2y = 3$
   $3x - 7y = 10$
   $[1, -1]$

3. $4a - 3b = -17$
   $5a + 8b = 14$
   $[-2, 3]$

4. $6x + 5y = -\frac{1}{3}$
   $3x - 2y = 2\frac{5}{6}$
   $[\frac{1}{2}, -\frac{2}{3}]$

5. $5x - 2y = 6$
   $10x - 4y = 14$

6. $2x - 3y = 5$
   $-6x + 9y = 1$

8.9   Square Matrices

UNIT MATRIX

Let **A** be a square matrix of order n. If **A** is pre- or post-multiplied by the unit matrix **I** of order n it is easily verified that the product in both cases is **A**, i.e.

$$\mathbf{IA} = \mathbf{AI} = \mathbf{A}.$$

DETERMINANT OF THE PRODUCT OF TWO SQUARE MATRICES

Let **A** and **B** be two square matrices of the same order. Since the two matrices are multiplied by the same rule as their determinants, it follows

$$|\mathbf{AB}| = |\mathbf{A}|\,|\mathbf{B}|$$

INVERSE OF A SQUARE MATRIX

Let **A** be a square matrix. If a matrix **B** exists such that

$$\mathbf{AB} = \mathbf{BA} = \mathbf{I}$$

then **B** is the inverse of the matrix **A**.

SINGULAR MATRIX

(*a*) A singular matrix has been defined as one whose determinant is zero. It may be shown that a singular matrix has no inverse. Let **A** be a singular matrix and suppose it has an inverse **B**, then

$$\mathbf{AB} = \mathbf{I} \quad \text{and} \quad |\mathbf{A}|\,|\mathbf{B}| = 1$$

But $|\mathbf{A}| = 0$, and the above equation is contradicted. Hence a singular matrix has no inverse.

(*b*) The inverse of a non-singular matrix is unique. Let **A** be a non-singular matrix and, if possible, let **B** and **C** both be inverses of **A**. Therefore

$$\mathbf{AB} = \mathbf{BA} = \mathbf{I}$$

and

$$AC = CA = I$$

Therefore

$$CAB = CI = C$$

and

$$CAB = IB = B$$

Therefore

$$B = C$$

We denote the inverse of a non-singular matrix **A** by **A**⁻¹ so that

$$AA^{-1} = A^{-1}A = I$$

8.10 The Transpose of a Matrix

If the rows and columns of an $(m \times n)$ matrix **A** are interchanged the resulting $(n \times m)$ matrix **A′** is said to be the transpose of **A**.

$$\text{If } A = \begin{pmatrix} 3 & -2 & 1 \\ 0 & 6 & 5 \\ -1 & 4 & -2 \\ 2 & 0 & -3 \end{pmatrix} \quad \text{then} \quad A' = \begin{pmatrix} 3 & 0 & -1 & 2 \\ -2 & 6 & 4 & 0 \\ 1 & 5 & -2 & -3 \end{pmatrix}$$

A square matrix **A** is said to be *symmetric* if **A′** = **A** and is said to be skew-symmetric if **A′** = −**A**, e.g.

$$A = \begin{pmatrix} a & h & g \\ h & b & f \\ g & f & c \end{pmatrix} \quad \text{is symmetric}$$

$$B = \begin{pmatrix} 0 & b & c \\ -b & 0 & d \\ -c & -d & 0 \end{pmatrix} \quad \text{is skew-symmetric}$$

Let us denote a non-singular square matrix $\mathbf{A}$ of order n by (a_{ij}) where i, j take the values $1, 2, \ldots, n$.

If $\mathbf{A}$ is symmetric, $\mathbf{A}' = (a_{ji})$ and $\mathbf{A}' = \mathbf{A}$. Thus $a_{ji} = a_{ij}$.

If $\mathbf{A}$ is skew-symmetric, $\mathbf{A}' = (a_{ji})$ and since $\mathbf{A}' = -\mathbf{A}$

$$(a_{ji}) = -(a_{ij}) \qquad \text{and thus} \qquad a_{ji} = -a_{ij}$$

On the leading diagonal, $i = j$. Therefore

$$a_{ii} = -a_{ii} = 0$$

so that all terms on the leading diagonal are zero. If $i \neq j$, $a_{ij} = -a_{ji}$.

8.11 Adjugate or Adjoint Matrix

If $\mathbf{A}$ is a square matrix, the *adjugate matrix* of $\mathbf{A}$ is defined by

$$\text{adj } \mathbf{A} = (\mathbf{A}_{ij})' = (\mathbf{A}_{ji})$$

i.e. adj $\mathbf{A}$ is formed by replacing each element of $\mathbf{A}$ by its cofactor in $|\mathbf{A}|$ and transposing the matrix so obtained. For example, the adjugate matrix of

$$\mathbf{A} = \begin{pmatrix} 1 & 2 & 1 \\ 0 & -3 & 4 \\ -2 & 1 & 0 \end{pmatrix}$$

is the transpose of

$$\begin{pmatrix} -4 & -8 & -6 \\ 1 & 2 & -5 \\ 11 & -4 & -3 \end{pmatrix}$$

i.e.

$$\text{adj } \mathbf{A} = \begin{pmatrix} -4 & 1 & 11 \\ -8 & 2 & -4 \\ -6 & -5 & -3 \end{pmatrix}$$

8.12   The Inverse of a Square Matrix

To show that

$$\mathbf{A}^{-1} = \frac{1}{|\mathbf{A}|} \text{ adj } \mathbf{A}$$

provided $|\mathbf{A}| \neq 0$, i.e. provided $\mathbf{A}$ is not a singular matrix, let

$$\mathbf{A} = \begin{pmatrix} a_{11} & a_{12} & \cdots & a_{1n} \\ a_{21} & a_{22} & \cdots & a_{2n} \\ \cdot & & & \\ \cdot & & & \\ \cdot & & & \\ a_{n1} & a_{n2} & \cdots & a_{nn} \end{pmatrix}$$

so that

$$\text{adj } \mathbf{A} = \begin{pmatrix} A_{11} & A_{21} & \cdots & A_{n1} \\ A_{12} & A_{22} & \cdots & A_{n2} \\ \cdot & & & \\ \cdot & & & \\ \cdot & & & \\ A_{1n} & A_{2n} & \cdots & A_{nn} \end{pmatrix}$$

Therefore

$$\mathbf{A}(\text{adj } \mathbf{A}) = (c_{ij})$$

where

$$c_{ij} = a_{i1}A_{j1} + a_{i2}A_{j2} + \cdots + a_{in}A_{jn} = \sum_{k=1}^{n} a_{ik}A_{jk}$$

If $i = j$, $c_{ij} = A$, and if $i \neq j$, $c_{ij} = 0$ (see Section 8.6), so that each of the elements on the leading diagonal is $|\mathbf{A}|$ and all other elements are zero.

Therefore, provided $|A| \neq 0$,

$$A \,(\text{adj } A) = \begin{pmatrix} |A| & 0 & 0 & \cdots & 0 \\ 0 & |A| & 0 & \cdots & 0 \\ 0 & 0 & |A| & \cdots & 0 \\ \vdots & & & & \\ 0 & 0 & 0 & \cdots & |A| \end{pmatrix}$$

$$= |A| \begin{pmatrix} 1 & 0 & 0 & \cdots & 0 \\ 0 & 1 & 0 & \cdots & 0 \\ 0 & 0 & 1 & \cdots & 0 \\ \vdots & & & & \\ 0 & 0 & 0 & \cdots & 1 \end{pmatrix}$$

$$= |A|I$$

(It may also be readily verified that $(\text{adj } A)A = |A|I$.)
It follows that

$$A\left(\frac{1}{|A|} \text{adj } A\right) = I, \text{ provided } |A| \neq 0$$

Hence

$$A^{-1} = \frac{1}{|A|} \text{adj } A$$

For example, if

$$A = \begin{pmatrix} 1 & 2 & 1 \\ 0 & -3 & 4 \\ -2 & 1 & 0 \end{pmatrix} \quad \text{then} \quad |A| = -26$$

$$A^{-1} = \frac{1}{|A|} \text{adj } A = -\frac{1}{26}\begin{pmatrix} -4 & 1 & 11 \\ -8 & 2 & -4 \\ -6 & -5 & -3 \end{pmatrix} \quad \text{(see Section 8.11)}$$

8.13 Solution of Linear Equations

Consider the set of n simultaneous linear equations in the n unknowns $x_1, x_2, \ldots, x_n$ as follows:

$$a_{i1}x_1 + a_{i2}x_2 + \cdots + a_{in}x_n = b_i \quad (i = 1, 2, \ldots, n)$$

These equations may be written in the matrix form

$$\mathbf{AX} = \mathbf{B}$$

where $\mathbf{A}$ is the $n \times n$ matrix of the coefficients a_{ij} and $\mathbf{X}$ and $\mathbf{B}$ are column matrices of the unknowns x_i and the quantities b_i respectively, each of order $n \times 1$.

If $\mathbf{A}$ is non-singular, then $\mathbf{A}^{-1}$ exists. On premultiplying each side of the above matrix equation by $\mathbf{A}^{-1}$, we have

$$\mathbf{A}^{-1}(\mathbf{AX}) = \mathbf{A}^{-1}\mathbf{B}$$

Now $\mathbf{A}^{-1}(\mathbf{AX}) = (\mathbf{A}^{-1}\mathbf{A})\mathbf{X} = \mathbf{IX} = \mathbf{X}$. Thus

$$\mathbf{X} = \mathbf{A}^{-1}\mathbf{B}$$

from which the solutions of the set of linear equations may be found.

EXAMPLE 8.3
Solve the following simultaneous equations.

(A) $x + 2y + z = -3$
$\quad\quad -3y + 4z = 17$
$\quad -2x + y = -5$

The coefficient matrix $\mathbf{A} = \begin{pmatrix} 1 & 2 & 1 \\ 0 & -3 & 4 \\ -2 & 1 & 0 \end{pmatrix}$.

In Section 8.12, we have shown that

$$\mathbf{A}^{-1} = \frac{1}{26}\begin{pmatrix} 4 & -1 & -11 \\ 8 & -2 & 4 \\ 6 & 5 & 3 \end{pmatrix}$$

Writing the equations in matrix form,

$$\begin{pmatrix} 1 & 2 & 1 \\ 0 & -3 & 4 \\ -2 & 1 & 0 \end{pmatrix}\begin{pmatrix} x \\ y \\ z \end{pmatrix} = \begin{pmatrix} -3 \\ 17 \\ -5 \end{pmatrix}$$

Multiplying both sides by $\mathbf{A}^{-1}$,

$$\frac{1}{26}\begin{pmatrix} 4 & -1 & -11 \\ 8 & -2 & 4 \\ 6 & 5 & 3 \end{pmatrix}\begin{pmatrix} 1 & 2 & 1 \\ 0 & -3 & 4 \\ -2 & 1 & 0 \end{pmatrix}\begin{pmatrix} x \\ y \\ z \end{pmatrix}$$

$$= \frac{1}{26}\begin{pmatrix} 4 & -1 & -11 \\ 8 & -2 & 4 \\ 6 & 5 & 3 \end{pmatrix}\begin{pmatrix} -3 \\ 17 \\ -5 \end{pmatrix}$$

$$\begin{pmatrix} x \\ y \\ z \end{pmatrix} = \frac{1}{26}\begin{pmatrix} 26 \\ -78 \\ 52 \end{pmatrix} = \begin{pmatrix} 1 \\ -3 \\ 2 \end{pmatrix}$$

$$x = 1 \qquad y = -3 \qquad z = 2$$

(B) $x + 2y + z = 1$

$3x - 2z = 0$

$2x + y = 0$

$$\begin{pmatrix} 1 & 2 & 1 \\ 3 & 0 & -2 \\ 2 & 1 & 0 \end{pmatrix}\begin{pmatrix} x \\ y \\ z \end{pmatrix} = \begin{pmatrix} 1 \\ 0 \\ 0 \end{pmatrix}$$

The cofactors of the first row of the coefficient matrix **A** are 2, −4, 3. These become the first column of adj **A**.

It may be verified that

$$\text{adj } A = \begin{pmatrix} 2 & 1 & -4 \\ -4 & -2 & 5 \\ 3 & 3 & -6 \end{pmatrix}$$

$$|A| = -3$$

$$\begin{pmatrix} x \\ y \\ z \end{pmatrix} = \begin{pmatrix} -\dfrac{2}{3} \\ \dfrac{4}{3} \\ -1 \end{pmatrix}$$

$$x = -\frac{2}{3} \qquad y = \frac{4}{3} \qquad z = -1$$

EXERCISES 8.5

1. Solve the following simultaneous equations:
   (i) The equations of Example 8.3(B) where 7, 8, 7 are substituted for 1, 0, 0, respectively on the right-hand sides.
   $$[2, 3, -1]$$
   (ii) $\quad x + y + z = 2$
   $$x - y + 5z = 4$$
   $$3x + y + 4z = -1$$
   $$[-6, 5, 3]$$
   (iii) $3a - 2b - d = 7$
   $$2b + 2c + d = 5$$
   $$a - 2b - 3c - 2d = -1$$
   $$b + 2c + d = 6$$
   $$[a = -10; b = -1; c = 21; d = -35]$$

2. For what value of c will the equations
   $$x + y - z = 3$$
   $$cx - y + 2z = 5$$
   $$x + 2y - z = 4$$
   have no unique solution.
   $$[c = -2]$$

9 Vectors and matrices

9.1 Column Vectors

In Cartesian coordinates, we may specify the position of a point P in the (x, y) plane by an ordered pair of numbers (a, b).

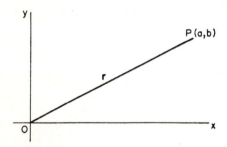

FIGURE 9.1

To reach the point P from the origin O, we move a units along Ox followed by b units parallel to Oy (Fig. 9.1). To move directly from O to P we move along the straight line OP.

$\overrightarrow{OP}$ is called the *position vector* of the point P and is usually represented by a column matrix or *column vector* as it is called.

$$\overrightarrow{OP} = \mathbf{r} = \begin{pmatrix} a \\ b \end{pmatrix} \qquad \text{a } (2 \times 1) \text{ matrix}$$

In three-dimensional coordinates, the position vector of the point (a, b, c) would be

$$\overrightarrow{OP} = \mathbf{r} = \begin{pmatrix} a \\ b \\ c \end{pmatrix} \qquad \text{a } (3 \times 1) \text{ matrix}$$

9.2 Addition of Two Vectors

Let the vectors be $\mathbf{r}_1 = \begin{pmatrix} 2 \\ 1 \end{pmatrix}$ and $\mathbf{r}_2 = \begin{pmatrix} 1 \\ 4 \end{pmatrix}$.

$$\mathbf{r}_1 + \mathbf{r}_2 = \begin{pmatrix} 2 + 1 \\ 1 + 4 \end{pmatrix} = \begin{pmatrix} 3 \\ 5 \end{pmatrix}$$

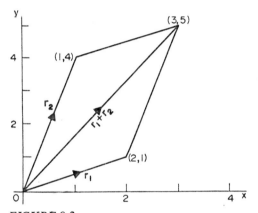

FIGURE 9.2

It will be seen (Fig. 9.2) that geometrically the vectors $\mathbf{r}_1$ and $\mathbf{r}_2$ are added by a parallelogram law.

9.3 Multiplication of a Vector by a Scalar

If the vector $\mathbf{r} = \begin{pmatrix} a \\ b \end{pmatrix}$ and k is a scalar, we define their product as

$$k\mathbf{r} = k\begin{pmatrix} a \\ b \end{pmatrix} = \begin{pmatrix} ka \\ kb \end{pmatrix}$$

$k\mathbf{r}$ has the same direction as $\mathbf{r}$ (Fig. 9.3) but its magnitude is increased by a factor k.

Further

$$\begin{pmatrix} k & 0 \\ 0 & k \end{pmatrix} \begin{pmatrix} a \\ b \end{pmatrix} = \begin{pmatrix} ka \\ kb \end{pmatrix}$$

i.e. multiplication by $\begin{pmatrix} k & 0 \\ 0 & k \end{pmatrix}$ is equivalent to multiplication by a scalar k

k and $\begin{pmatrix} k & 0 \\ 0 & k \end{pmatrix}$ is called a *scalar matrix*.

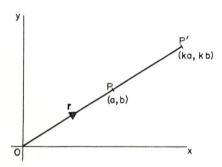

FIGURE 9.3

9.4   Translation

Apply a translation to a plane such that every point is carried 2 units parallel to Ox and -1 unit parallel to Oy.

It is clear from Fig. 9.4, that any point $Q(x, y)$ is carried to $Q'(x + 2, y - 1)$.

This is equivalent to

$$\begin{pmatrix} x \\ y \end{pmatrix} + \begin{pmatrix} 2 \\ -1 \end{pmatrix} = \begin{pmatrix} x + 2 \\ y - 1 \end{pmatrix}$$

$\begin{pmatrix} 2 \\ -1 \end{pmatrix}$ is called the *shift vector* of the translation.

The origin will be carried to P, whose position vector is $\begin{pmatrix} 2 \\ -1 \end{pmatrix}$, i.e. $\overrightarrow{OP}$ is the shift vector of the translation.

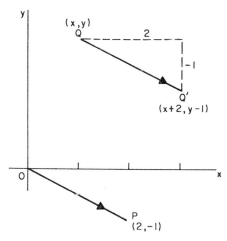

FIGURE 9.4

9.5   Base Vectors

The vectors $\begin{pmatrix} 1 \\ 0 \end{pmatrix}$ and $\begin{pmatrix} 0 \\ 1 \end{pmatrix}$ are shifts of unit length in the directions Ox and Oy respectively.

These two vectors are called *base vectors* because any vector may be expressed in terms of them as follows:

$$\begin{pmatrix} a \\ b \end{pmatrix} = \begin{pmatrix} a \\ 0 \end{pmatrix} + \begin{pmatrix} 0 \\ b \end{pmatrix} = a\begin{pmatrix} 1 \\ 0 \end{pmatrix} + b\begin{pmatrix} 0 \\ 1 \end{pmatrix}$$

If we write

$$\mathbf{i} = \begin{pmatrix} 1 \\ 0 \end{pmatrix} \quad \text{and} \quad \mathbf{j} = \begin{pmatrix} 0 \\ 1 \end{pmatrix}$$

the position vector of a point P (Fig. 9.5) may be written

$$\overrightarrow{OP} = \mathbf{r} = \begin{pmatrix} a \\ b \end{pmatrix} = a\mathbf{i} + b\mathbf{j} \tag{9.1}$$

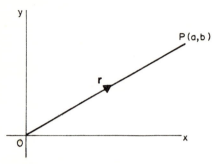

FIGURE 9.5

9.6   Elementary Matrices

The matrices

$$\begin{pmatrix} 0 & 1 \\ 1 & 0 \end{pmatrix} \quad \begin{pmatrix} k & 0 \\ 0 & 1 \end{pmatrix} \quad \begin{pmatrix} 1 & 0 \\ 0 & k \end{pmatrix} \quad \begin{pmatrix} 1 & 0 \\ k & 1 \end{pmatrix} \quad \begin{pmatrix} 1 & k \\ 0 & 1 \end{pmatrix}$$

are called *elementary matrices, k* being a scalar (real number). Multiplication by these elementary matrices leads to results which are analogues of those which arise in the manipulation of determinants.

$$\begin{pmatrix} 0 & 1 \\ 1 & 0 \end{pmatrix}\begin{pmatrix} a & b \\ c & d \end{pmatrix} = \begin{pmatrix} c & d \\ a & b \end{pmatrix} \qquad \text{Interchange of rows of matrix } \begin{pmatrix} a & b \\ c & d \end{pmatrix}$$

$$\begin{pmatrix} k & 0 \\ 0 & 1 \end{pmatrix}\begin{pmatrix} a & b \\ c & d \end{pmatrix} = \begin{pmatrix} ka & kb \\ c & d \end{pmatrix} \qquad \text{1st row multiplied by } k$$

$$\begin{pmatrix} 1 & 0 \\ 0 & k \end{pmatrix}\begin{pmatrix} a & b \\ c & d \end{pmatrix} = \begin{pmatrix} a & b \\ kc & kd \end{pmatrix} \qquad \text{2nd row multiplied by } k$$

$$\begin{pmatrix} 1 & k \\ 0 & 1 \end{pmatrix}\begin{pmatrix} a & b \\ c & d \end{pmatrix} = \begin{pmatrix} a + kc & b + kd \\ c & d \end{pmatrix} \qquad \text{Multiple of 2nd row added to the 1st row}$$

$$\begin{pmatrix} 1 & 0 \\ k & 1 \end{pmatrix}\begin{pmatrix} a & b \\ c & d \end{pmatrix} = \begin{pmatrix} a & b \\ c + ka & d + kb \end{pmatrix} \qquad \text{Multiple of 1st row added to the 2nd row}$$

9.7   Linear Transformations of a Plane

Various transformations of the plane, such as reflections, rotations, etc., may be expressed in terms of elementary (2 × 2) matrices.

Consider the simplest transformation—reflection in O*x* (Fig. 9.6). P(*x, y*) is transformed into P′(*x′, y′*) such that

$$x' = x \qquad y' = -y$$

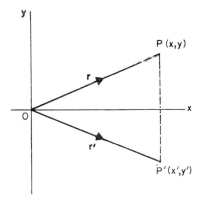

FIGURE 9.6

r, the position vector of P, may be expressed as a column vector $\begin{pmatrix} x \\ y \end{pmatrix}$.

r′, the position vector of P′, may be expressed as a column vector $\begin{pmatrix} x' \\ y' \end{pmatrix}$.

The equations of the transformation may be expressed in matrix notation thus:

$$\begin{pmatrix} x' \\ y' \end{pmatrix} = \begin{pmatrix} 1 & 0 \\ 0 & -1 \end{pmatrix}\begin{pmatrix} x \\ y \end{pmatrix} = \begin{pmatrix} x \\ -y \end{pmatrix}$$

i.e. by the matrix equation

$$\mathbf{r'} = \mathbf{Ar} \qquad \text{where } \mathbf{A} = \begin{pmatrix} 1 & 0 \\ 0 & -1 \end{pmatrix}$$

A is the matrix of the transformation of reflection.

In general, if any column vector is premultiplied by a (2 × 2) matrix it is *transformed* or *mapped* into another column vector.

If $\mathbf{A} = \begin{pmatrix} a & b \\ c & d \end{pmatrix}$ is the matrix of the transformation, then

$$\begin{pmatrix} x' \\ y' \end{pmatrix} = \begin{pmatrix} a & b \\ c & d \end{pmatrix} \begin{pmatrix} x \\ y \end{pmatrix}$$

i.e.

$$\begin{pmatrix} x' \\ y' \end{pmatrix} = \mathbf{A} \begin{pmatrix} x \\ y \end{pmatrix}$$

or

$$\mathbf{r}' = \mathbf{A}\mathbf{r} \tag{9.2}$$

and

$$x' = ax + by$$
$$y' = cx + dy$$

Thus the transformation is defined by linear functions of x and y.

Note that for all such transformations $(0, 0)$ is transformed into $(0, 0)$, i.e. they keep the origin fixed.

9.8 Transformation of Base Vectors

$$\begin{pmatrix} a & b \\ c & d \end{pmatrix} \begin{pmatrix} 1 \\ 0 \end{pmatrix} = \begin{pmatrix} a \\ c \end{pmatrix} \quad \text{i.e. } \mathbf{A}\mathbf{i} = \begin{pmatrix} a \\ c \end{pmatrix} = \mathbf{i}' \text{ (say)}$$

$$\begin{pmatrix} a & b \\ c & d \end{pmatrix} \begin{pmatrix} 0 \\ 1 \end{pmatrix} = \begin{pmatrix} b \\ d \end{pmatrix} \quad \text{i.e. } \mathbf{A}\mathbf{j} = \begin{pmatrix} b \\ d \end{pmatrix} = \mathbf{j}' \text{ (say)}$$

The base vectors are transformed into vectors $\mathbf{i}'$, $\mathbf{j}'$ which are represented by the columns of $\mathbf{A}$.

This means that, if the images of the points $(1, 0)$, $(0, 1)$ under a certain transformation are known, the transformation is known.

$$\begin{pmatrix} x' \\ y' \end{pmatrix} = \begin{pmatrix} a & b \\ c & d \end{pmatrix}\begin{pmatrix} x \\ y \end{pmatrix} = \begin{pmatrix} ax + by \\ cx + dy \end{pmatrix} = \begin{pmatrix} ax \\ cx \end{pmatrix} + \begin{pmatrix} by \\ dy \end{pmatrix} = x\begin{pmatrix} a \\ c \end{pmatrix} + y\begin{pmatrix} b \\ d \end{pmatrix}$$

(9.3)

Therefore

$$A(x\mathbf{i} + y\mathbf{j}) = x\mathbf{Ai} + y\mathbf{Aj} = x\mathbf{i}' + y\mathbf{j}'$$

(9.4)

This means that the point (x, y) maps on to a point whose position vector is the same combination of $\mathbf{i}'$, $\mathbf{j}'$ as the position vector of (x, y) is of $\mathbf{i}$, $\mathbf{j}$.
Let us consider the effect of applying the transformation

$$\mathbf{M} = \begin{pmatrix} 4 & 1 \\ 3 & 2 \end{pmatrix}$$

to the plane.
How is the space OACB (a unit square) transformed? Using the symbol $\rightarrow$ to represent "is transformed into",

$$\begin{pmatrix} 1 \\ 0 \end{pmatrix} \rightarrow \begin{pmatrix} 4 \\ 3 \end{pmatrix} = \mathbf{i}'$$

i.e. the point $A(1, 0)$ is mapped on to the point $A'(4, 3)$.

$$\begin{pmatrix} 0 \\ 1 \end{pmatrix} \rightarrow \begin{pmatrix} 1 \\ 2 \end{pmatrix} = \mathbf{j}'$$

i.e. the point $B(0, 1)$ is mapped on to the point $B'(1, 2)$.
Any point (x, y) is mapped on to $(4x + y, 3x + 2y)$ since

$$\begin{pmatrix} x' \\ y' \end{pmatrix} = x\begin{pmatrix} 4 \\ 3 \end{pmatrix} + y\begin{pmatrix} 1 \\ 2 \end{pmatrix} \qquad \text{by (9.3)}$$

In particular, the point $C(1, 1)$ is mapped on to the point $C'(5, 5)$. The square OACB is thus mapped on to OA$'$C$'$B$'$ which is clearly a parallelogram (Fig. 9.7).

The point $(x, 0) \rightarrow (4x, 3x)$ i.e. $x(4, 3)$

The point $(0, y) \rightarrow (y, 2y)$ i.e. $y(1, 2)$

so that, for example, the square ODFE (Fig. 9.8(a)) will transform into the parallelogram OD$'$F$'$E$'$ (Fig. 9.8(b)).

It follows that the grid of squares of unit side will transform into a grid of parallelograms congruent to OA$'$C$'$B$'$.

Note: $\mathbf{N} = \begin{pmatrix} 1 & 2 \\ 3 & 6 \end{pmatrix}$ is a singular matrix since $|\mathbf{N}| = 0$.

Under this transformation $\begin{pmatrix} x' \\ y' \end{pmatrix} = x \begin{pmatrix} 1 \\ 3 \end{pmatrix} + y \begin{pmatrix} 2 \\ 6 \end{pmatrix}$

i.e. $x' = x + 2y$ and $y' = 3x + 6y$

Therefore $y' = 3x'$.

All points in the plane are transformed on to points lying on one straight line.

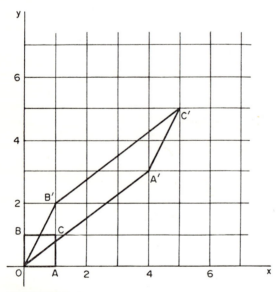

FIGURE 9.7

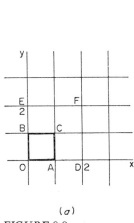
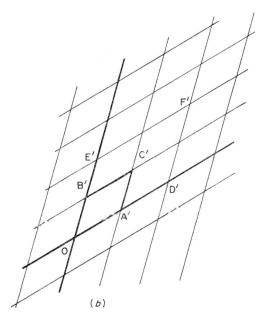

(a)   (b)

FIGURE 9.8

EXERCISES 9.1

Using graph paper, show how the unit square OACB (Fig. 9.7) is transformed under the transformations represented by the following matrices:

(i) $\begin{pmatrix} 3 & 0 \\ 0 & 3 \end{pmatrix}$ (ii) $\begin{pmatrix} 4 & 2 \\ -1 & 1 \end{pmatrix}$ (iii) $\begin{pmatrix} 1 & 2 \\ 0 & 1 \end{pmatrix}$ (shear)   (iv) $\begin{pmatrix} 2 & -1 \\ 1 & 2 \end{pmatrix}$ (v) $\begin{pmatrix} 3 & -6 \\ -2 & 4 \end{pmatrix}$

9.9 Affine Transformations

If the matrix $A = \begin{pmatrix} a & b \\ c & d \end{pmatrix}$ is non-singular, a grid of squares (Fig. 9.8(a)) is mapped on to a grid of parallelograms (Fig. 9.8(b)) such that

(i) parallel lines remain parallel
(ii) the ratios of lengths in the same direction remain unchanged.

Such transformations are called *affine*.

If the parallelogram OA′C′B′, into which the unit square OACB is transformed, is known, the transformation matrix A is known.

9.10   Eight Simple Transformations

1. REFLECTION IN Ox (Fig. 9.9)

Under reflection in Ox,

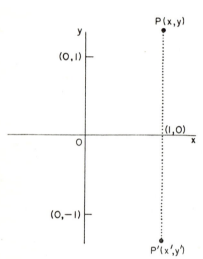

the base vector $\begin{pmatrix} 1 \\ 0 \end{pmatrix}$ is transformed into $\begin{pmatrix} 1 \\ 0 \end{pmatrix}$ and the base vector $\begin{pmatrix} 0 \\ 1 \end{pmatrix}$ into $\begin{pmatrix} 0 \\ -1 \end{pmatrix}$

By Section 9.8, the appropriate transformation matrix is

$$\mathbf{A}_x = \begin{pmatrix} 1 & 0 \\ 0 & -1 \end{pmatrix}$$

$$\begin{pmatrix} x' \\ y' \end{pmatrix} = \begin{pmatrix} 1 & 0 \\ 0 & -1 \end{pmatrix}\begin{pmatrix} x \\ y \end{pmatrix} = \begin{pmatrix} x \\ -y \end{pmatrix}$$

FIGURE 9.9

Therefore

$$x' = x$$
$$y' = -y$$

2. REFLECTION IN Oy (Fig. 9.10)

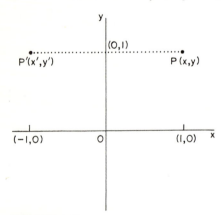

In this case

$$\begin{pmatrix} 1 \\ 0 \end{pmatrix} \rightarrow \begin{pmatrix} -1 \\ 0 \end{pmatrix} \qquad \begin{pmatrix} 0 \\ 1 \end{pmatrix} \rightarrow \begin{pmatrix} 0 \\ 1 \end{pmatrix}$$

The required transformation matrix is

$$\mathbf{A}_y = \begin{pmatrix} -1 & 0 \\ 0 & 1 \end{pmatrix}$$

$$\begin{pmatrix} x' \\ y' \end{pmatrix} = \begin{pmatrix} -1 & 0 \\ 0 & 1 \end{pmatrix}\begin{pmatrix} x \\ y \end{pmatrix} = \begin{pmatrix} -x \\ y \end{pmatrix}$$

$$x' = -x$$
$$y = y$$

FIGURE 9.10

3. REFLECTION IN THE LINE $x = y$ (Fig. 9.11)

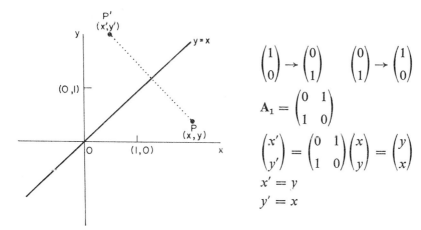

$$\begin{pmatrix} 1 \\ 0 \end{pmatrix} \rightarrow \begin{pmatrix} 0 \\ 1 \end{pmatrix} \qquad \begin{pmatrix} 0 \\ 1 \end{pmatrix} \rightarrow \begin{pmatrix} 1 \\ 0 \end{pmatrix}$$

$$\mathbf{A}_1 = \begin{pmatrix} 0 & 1 \\ 1 & 0 \end{pmatrix}$$

$$\begin{pmatrix} x' \\ y' \end{pmatrix} = \begin{pmatrix} 0 & 1 \\ 1 & 0 \end{pmatrix} \begin{pmatrix} x \\ y \end{pmatrix} = \begin{pmatrix} y \\ x \end{pmatrix}$$

$$x' = y$$
$$y' = x$$

FIGURE 9.11

4. REFLECTION IN THE LINE $y = -x$ (Fig. 9.12)

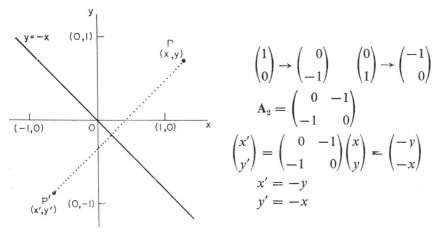

$$\begin{pmatrix} 1 \\ 0 \end{pmatrix} \rightarrow \begin{pmatrix} 0 \\ -1 \end{pmatrix} \qquad \begin{pmatrix} 0 \\ 1 \end{pmatrix} \rightarrow \begin{pmatrix} -1 \\ 0 \end{pmatrix}$$

$$\mathbf{A}_2 = \begin{pmatrix} 0 & -1 \\ -1 & 0 \end{pmatrix}$$

$$\begin{pmatrix} x' \\ y' \end{pmatrix} = \begin{pmatrix} 0 & -1 \\ -1 & 0 \end{pmatrix} \begin{pmatrix} x \\ y \end{pmatrix} = \begin{pmatrix} -y \\ -x \end{pmatrix}$$

$$x' = -y$$
$$y' = -x$$

FIGURE 9.12

From geometrical considerations, we should expect the matrices representing reflections to be their own inverses. This is easily verified by direct multiplication:

$$\mathbf{A_2} \times \mathbf{A_2} = \begin{pmatrix} 0 & -1 \\ -1 & 0 \end{pmatrix}\begin{pmatrix} 0 & -1 \\ -1 & 0 \end{pmatrix} = \begin{pmatrix} 1 & 0 \\ 0 & 1 \end{pmatrix} = \mathbf{I}$$

5. QUARTER TURN ABOUT THE ORIGIN (Fig. 9.13)

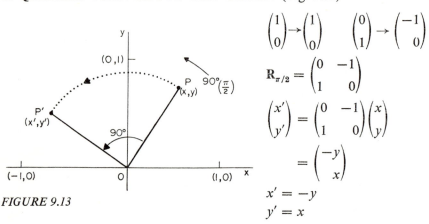

$$\begin{pmatrix} 1 \\ 0 \end{pmatrix} \rightarrow \begin{pmatrix} 1 \\ 0 \end{pmatrix} \qquad \begin{pmatrix} 0 \\ 1 \end{pmatrix} \rightarrow \begin{pmatrix} -1 \\ 0 \end{pmatrix}$$

$$\mathbf{R}_{\pi/2} = \begin{pmatrix} 0 & -1 \\ 1 & 0 \end{pmatrix}$$

$$\begin{pmatrix} x' \\ y' \end{pmatrix} = \begin{pmatrix} 0 & -1 \\ 1 & 0 \end{pmatrix}\begin{pmatrix} x \\ y \end{pmatrix}$$

$$= \begin{pmatrix} -y \\ x \end{pmatrix}$$

$$x' = -y$$
$$y' = x$$

FIGURE 9.13

6. HALF TURN ABOUT THE ORIGIN (Fig. 9.14)

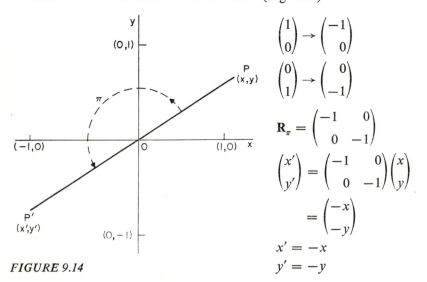

$$\begin{pmatrix} 1 \\ 0 \end{pmatrix} \rightarrow \begin{pmatrix} -1 \\ 0 \end{pmatrix}$$

$$\begin{pmatrix} 0 \\ 1 \end{pmatrix} \rightarrow \begin{pmatrix} 0 \\ -1 \end{pmatrix}$$

$$\mathbf{R}_{\pi} = \begin{pmatrix} -1 & 0 \\ 0 & -1 \end{pmatrix}$$

$$\begin{pmatrix} x' \\ y' \end{pmatrix} = \begin{pmatrix} -1 & 0 \\ 0 & -1 \end{pmatrix}\begin{pmatrix} x \\ y \end{pmatrix}$$

$$= \begin{pmatrix} -x \\ -y \end{pmatrix}$$

$$x' = -x$$
$$y' = -y$$

FIGURE 9.14

7. THREE-QUARTER TURN ABOUT THE ORIGIN (Fig. 9.15)

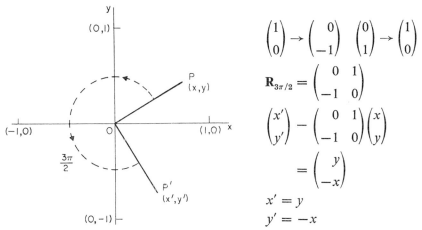

$$\begin{pmatrix} 1 \\ 0 \end{pmatrix} \rightarrow \begin{pmatrix} 0 \\ -1 \end{pmatrix} \quad \begin{pmatrix} 0 \\ 1 \end{pmatrix} \rightarrow \begin{pmatrix} 1 \\ 0 \end{pmatrix}$$

$$\mathbf{R}_{3\pi/2} = \begin{pmatrix} 0 & 1 \\ -1 & 0 \end{pmatrix}$$

$$\begin{pmatrix} x' \\ y' \end{pmatrix} - \begin{pmatrix} 0 & 1 \\ -1 & 0 \end{pmatrix}\begin{pmatrix} x \\ y \end{pmatrix}$$

$$= \begin{pmatrix} y \\ -x \end{pmatrix}$$

$$x' = y$$
$$y' = -x$$

FIGURE 9.15

8. THE IDENTITY TRANSFORMATION

$$\begin{pmatrix} 1 \\ 0 \end{pmatrix} \rightarrow \begin{pmatrix} 1 \\ 0 \end{pmatrix} \quad \begin{pmatrix} 0 \\ 1 \end{pmatrix} \rightarrow \begin{pmatrix} 0 \\ 1 \end{pmatrix}$$

$$\mathbf{I} = \begin{pmatrix} 1 & 0 \\ 0 & 1 \end{pmatrix}$$

$$\begin{pmatrix} x' \\ y' \end{pmatrix} = \begin{pmatrix} 1 & 0 \\ 0 & 1 \end{pmatrix}\begin{pmatrix} x \\ y \end{pmatrix} = \begin{pmatrix} x \\ y \end{pmatrix}$$

$$x' = x \quad y' = y$$

The identity transformation leaves all points unchanged.

EXAMPLE 9.1
(A) From geometrical considerations
    (i) the half turn matrix is its own inverse,
    (ii) the quarter turn matrix is the inverse of a three-quarter turn matrix and
        vice versa,
since the effect of applying two half turns or a quarter turn and a three-quarter turn in succession is to return all points in the plane to their initial positions.

This may be verified by direct multiplication of the respective matrices

(i) $R_\pi \times R_\pi = \begin{pmatrix} -1 & 0 \\ 0 & -1 \end{pmatrix} \begin{pmatrix} -1 & 0 \\ 0 & -1 \end{pmatrix} = \begin{pmatrix} 1 & 0 \\ 0 & 1 \end{pmatrix} = I$

(ii) $R_{\pi/2} \times R_{3\pi/2} = \begin{pmatrix} 0 & -1 \\ 1 & 0 \end{pmatrix} \begin{pmatrix} 0 & 1 \\ -1 & 0 \end{pmatrix} = \begin{pmatrix} 1 & 0 \\ 0 & 1 \end{pmatrix} = I$

(B) Verify that a reflection in the line $y = -x$ followed by a half turn is equivalent to a reflection in the line $y = x$.

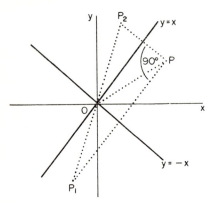

FIGURE 9.16

P_1 is the image of P under reflection in $y = -x$, i.e. under A_2 (Fig. 9.16).
P_2 is the image of P_1 under a half turn, i.e. under R_π.
Now

$$R_\pi \times A_2 = \begin{pmatrix} -1 & 0 \\ 0 & -1 \end{pmatrix} \begin{pmatrix} 0 & -1 \\ -1 & 0 \end{pmatrix} = \begin{pmatrix} 0 & 1 \\ 1 & 0 \end{pmatrix} = A_1$$

Thus P_2 is the image of P under reflection in $y = -x$.
We deduce the following geometrical property of any right-angled triangle: If O is the mid-point of the hypotenuse P_1P_2 of the right-angled triangle P_1P_2P, then

$OP = OP_1 = OP_2$

EXERCISES 9.2
1. The point (x, y) is mapped onto the point (x', y') by the rule

$$\begin{pmatrix} x' \\ y \end{pmatrix} = \begin{pmatrix} 3 & 0 \\ 0 & 3 \end{pmatrix} \begin{pmatrix} x \\ y \end{pmatrix}$$

What are the equations giving (x', y') in terms of (x, y)?
Describe this transformation.

2. What is the product of the matrices $\begin{pmatrix} 1 & 0 \\ 0 & 0 \end{pmatrix}$ and $\begin{pmatrix} x \\ y \end{pmatrix}$?

   Describe the transformation given by the matrix $\begin{pmatrix} 1 & 0 \\ 0 & 0 \end{pmatrix}$.

3. A transformation with matrix $M = \begin{pmatrix} 0 & -1 \\ 1 & 0 \end{pmatrix}$ takes the square OABC where O
   is $(0, 0)$, A is $(1, 0)$, B is $(1, 1)$ and C is $(0, 1)$ into the position OA′B′C′ Give
   the coordinates of A′, B′, C′ and show OABC and OA′B′C′ on the same piece of
   graph paper.

4. Answer question 3 if (i) $M = \begin{pmatrix} 1 & 2 \\ 0 & 1 \end{pmatrix}$, (ii) $M = \begin{pmatrix} -2 & 0 \\ 0 & -2 \end{pmatrix}$, and describe
   these transformations.

5. What single transformation is equivalent to the transformation A_y followed by
   A_x? Illustrate by a diagram.
   Does $A_x A_y = A_y A_x$?
   $[R_\pi; \text{yes}]$

6. Show that a reflection in Ox followed by a reflection in the line $y = x$ is equivalent
   to a quarter turn. What is the result of applying these transformations in the
   reverse order?
   $[R_{3\pi/2}]$

9.11 Further Transformations

1. ENLARGEMENT (Fig. 9.17)

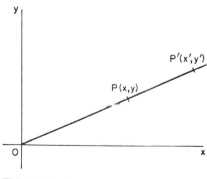

FIGURE 9.17

Let the factor of enlargement be k,
so that $OP' = kOP$.

$$\begin{pmatrix} 1 \\ 0 \end{pmatrix} \rightarrow \begin{pmatrix} k \\ 0 \end{pmatrix} \qquad \begin{pmatrix} 0 \\ 1 \end{pmatrix} \rightarrow \begin{pmatrix} 0 \\ k \end{pmatrix}$$

$$A_e = \begin{pmatrix} k & 0 \\ 0 & k \end{pmatrix}$$

$$\begin{pmatrix} x' \\ y' \end{pmatrix} = \begin{pmatrix} k & 0 \\ 0 & k \end{pmatrix} \begin{pmatrix} x \\ y \end{pmatrix} = \begin{pmatrix} kx \\ ky \end{pmatrix}$$

$$x' = kx$$
$$y' = ky$$

To invert an enlargement of magnification k, we must apply an enlargement of magnification $1/k$ represented by the matrix

$$\begin{pmatrix} 1/k & 0 \\ 0 & 1/k \end{pmatrix}$$

which must be the inverse of the matrix $\begin{pmatrix} k & 0 \\ 0 & k \end{pmatrix}$.

2. SHEARING PARALLEL TO Ox (Fig. 9.18)

Let the shear be such that $(0, 1)$ becomes $(k, 1)$. Therefore

$$\begin{pmatrix} 1 \\ 0 \end{pmatrix} \rightarrow \begin{pmatrix} 1 \\ 0 \end{pmatrix} \qquad \begin{pmatrix} 0 \\ 1 \end{pmatrix} \rightarrow \begin{pmatrix} k \\ 1 \end{pmatrix}$$

$$S_x = \begin{pmatrix} 1 & k \\ 0 & 1 \end{pmatrix}$$

$$\begin{pmatrix} x' \\ y' \end{pmatrix} = \begin{pmatrix} 1 & k \\ 0 & 1 \end{pmatrix}\begin{pmatrix} x \\ y \end{pmatrix} = \begin{pmatrix} x + ky \\ y \end{pmatrix}$$

$$x' = x + ky \qquad y' = y$$

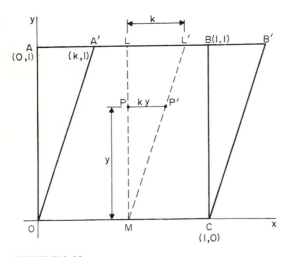

FIGURE 9.18

The unit square OABC is mapped on the parallelogram OA′R′C.

Any point P(x, y) is mapped onto P′ such that PP′ is horizontal and of length ky. All points on the line LPM move horizontally onto the line L′P′M.

The inverse transformation will restore (k, 1) to (0, 1) and will take (0, 1) to ($-k$, 1) whilst (1, 0) is unchanged. The inverse matrix is therefore $\begin{pmatrix} 1 & -k \\ 0 & 1 \end{pmatrix}$.

For shearing parallel to Oy, the transformation matrix is

$$S_y = \begin{pmatrix} 1 & 0 \\ k & 1 \end{pmatrix} \quad \text{and its inverse is} \quad \begin{pmatrix} 1 & 0 \\ -k & 1 \end{pmatrix}$$

3. ROTATIONS (Fig. 9.19)

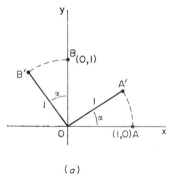 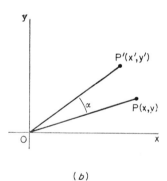

(a)   (b)

FIGURE 9.19

Consider first rotation through an angle α.

$$\begin{pmatrix} 1 \\ 0 \end{pmatrix} \to \begin{pmatrix} \cos \alpha \\ \sin \alpha \end{pmatrix} \quad \begin{pmatrix} 0 \\ 1 \end{pmatrix} \to \begin{pmatrix} -\sin \alpha \\ \cos \alpha \end{pmatrix}$$

The transformation matrix is therefore

$$R_\alpha = \begin{pmatrix} \cos \alpha & -\sin \alpha \\ \sin \alpha & \cos \alpha \end{pmatrix}$$

$$\begin{pmatrix} x' \\ y' \end{pmatrix} = \begin{pmatrix} \cos \alpha & -\sin \alpha \\ \sin \alpha & \cos \alpha \end{pmatrix} \begin{pmatrix} x \\ y \end{pmatrix}$$

$$x' = x \cos \alpha - y \sin \alpha$$

$$y' = x \sin \alpha + y \cos \alpha$$

If the first rotation α is followed by a second rotation β so that

$$\begin{pmatrix} x' \\ y' \end{pmatrix} \rightarrow \begin{pmatrix} x'' \\ y'' \end{pmatrix}$$

$$\begin{pmatrix} x'' \\ y'' \end{pmatrix} = \begin{pmatrix} \cos \beta & -\sin \beta \\ \sin \beta & \cos \beta \end{pmatrix} \begin{pmatrix} x' \\ y' \end{pmatrix}$$

$$= \begin{pmatrix} \cos \beta & -\sin \beta \\ \sin \beta & \cos \beta \end{pmatrix} \begin{pmatrix} \cos \alpha & -\sin \alpha \\ \sin \alpha & \cos \alpha \end{pmatrix} \begin{pmatrix} x \\ y \end{pmatrix}$$

$$\begin{pmatrix} x'' \\ y'' \end{pmatrix} = \begin{pmatrix} (\cos \alpha \cos \beta - \sin \alpha \sin \beta) & -(\sin \alpha \cos \beta + \cos \alpha \sin \beta) \\ (\cos \alpha \sin \beta + \sin \alpha \cos \beta) & (\cos \alpha \cos \beta - \sin \alpha \sin \beta) \end{pmatrix} \begin{pmatrix} x \\ y \end{pmatrix}$$

But since a rotation α followed by a rotation β is equivalent to a rotation $(\alpha + \beta)$,

$$\begin{pmatrix} x'' \\ y'' \end{pmatrix} = \begin{pmatrix} \cos (\alpha + \beta) & -\sin (\alpha + \beta) \\ \sin (\alpha + \beta) & \cos (\alpha + \beta) \end{pmatrix} \begin{pmatrix} x \\ y \end{pmatrix}$$

Therefore

$$\sin (\alpha + \beta) = \sin \alpha \cos \beta + \cos \alpha \sin \beta$$
$$\cos (\alpha + \beta) = \cos \alpha \cos \beta - \sin \alpha \sin \beta$$

4. REFLECTION IN THE LINE $y = x \tan \alpha$ (Fig. 9.20)

$$\begin{pmatrix} 1 \\ 0 \end{pmatrix} \rightarrow \begin{pmatrix} \cos 2\alpha \\ \sin 2\alpha \end{pmatrix} \qquad \begin{pmatrix} 0 \\ 1 \end{pmatrix} \rightarrow \begin{pmatrix} \sin 2\alpha \\ -\cos 2\alpha \end{pmatrix}$$

$$\mathbf{A}_\alpha = \begin{pmatrix} \cos 2\alpha & \sin 2\alpha \\ \sin 2\alpha & -\cos 2\alpha \end{pmatrix}$$

It is interesting to see how this matrix may be derived by applying a succession of transformations.

(i) First apply the transformation $\mathbf{R}_{-\alpha}$ (corresponding to rotation through angle α clockwise) which maps $y = x \tan \alpha$ on Ox.

Now apply the transformation for reflection in Ox, i.e. $\mathbf{A}_x$, and finally apply the transformation corresponding to rotation α in the anticlockwise

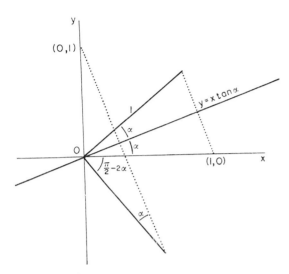

FIGURE 9.20

direction $\mathbf{R}_\alpha$. These three transformations are equivalent to reflection in $y = x \tan \alpha$. Therefore

$$\mathbf{A}_\alpha = \mathbf{R}_\alpha \mathbf{A}_x \mathbf{R}_{-\alpha}$$

$$= \begin{pmatrix} \cos \alpha & -\sin \alpha \\ \sin \alpha & \cos \alpha \end{pmatrix} \begin{pmatrix} 1 & 0 \\ 0 & -1 \end{pmatrix} \begin{pmatrix} \cos \alpha & \sin \alpha \\ -\sin \alpha & \cos \alpha \end{pmatrix}$$

$$= \begin{pmatrix} \cos \alpha & -\sin \alpha \\ \sin \alpha & \cos \alpha \end{pmatrix} \begin{pmatrix} \cos \alpha & \sin \alpha \\ \sin \alpha & -\cos \alpha \end{pmatrix}$$

$$= \begin{pmatrix} (\cos^2 \alpha - \sin^2 \alpha) & (2 \sin \alpha \, \cos \alpha) \\ (2 \sin \alpha \, \cos \alpha) & (\sin^2 \alpha - \cos^2 \alpha) \end{pmatrix}$$

$$= \begin{pmatrix} \cos 2\alpha & \sin 2\alpha \\ \sin 2\alpha & -\cos 2\alpha \end{pmatrix} \quad \text{as above}$$

(ii) Reflection in Ox ($\mathbf{A}_x$) followed by reflection in $y = x \tan \alpha$ ($\mathbf{A}_\alpha$) is equivalent to rotation through 2α ($\mathbf{R}_{2\alpha}$) (Fig. 9.21). Therefore

$$\mathbf{A}_\alpha \mathbf{A}_x = \mathbf{R}_{2\alpha}$$

$$\mathbf{A}_\alpha \mathbf{A}_x \mathbf{A}_x^{-1} = \mathbf{R}_{2\alpha} \mathbf{A}_x^{-1}$$

$$\mathbf{A}_\alpha = \mathbf{R}_{2\alpha} \mathbf{A}_x^{-1}$$

Therefore

$$\mathbf{A}_\alpha = \begin{pmatrix} \cos 2\alpha & -\sin 2\alpha \\ \sin 2\alpha & \cos 2\alpha \end{pmatrix} \begin{pmatrix} 1 & 0 \\ 0 & -1 \end{pmatrix}$$

$$= \begin{pmatrix} \cos 2\alpha & \sin 2\alpha \\ \sin 2\alpha & -\cos 2\alpha \end{pmatrix} \quad \text{as before}$$

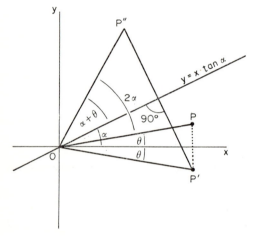

FIGURE 9.21

9.12 *Eigenvectors*

Let $\mathbf{A} = \begin{pmatrix} a & b \\ c & d \end{pmatrix}$ be a non-singular matrix.

Consider the transformation

$$\begin{pmatrix} a & b \\ c & d \end{pmatrix} \begin{pmatrix} x \\ y \end{pmatrix} = \begin{pmatrix} x' \\ y' \end{pmatrix}$$

Column vectors $\begin{pmatrix} x \\ y \end{pmatrix}$ whose directions are unchanged by this transformation are called *eigenvectors* of the transformation.

A trivial case is that of enlargement

$$\begin{pmatrix} k & 0 \\ 0 & k \end{pmatrix} \begin{pmatrix} x \\ y \end{pmatrix} = \begin{pmatrix} kx \\ ky \end{pmatrix}$$

in which the direction of *every* vector remains unchanged.
    In the more general case we must investigate the transformation

$$\begin{pmatrix} a & b \\ c & d \end{pmatrix} \begin{pmatrix} x \\ y \end{pmatrix} = \begin{pmatrix} kx \\ ky \end{pmatrix} \tag{9.5}$$

for which the directions of some vectors are unchanged.
    Consider the following example. What vectors have their directions unchanged under the transformation $\begin{pmatrix} 3 & -5 \\ 2 & -4 \end{pmatrix}$?

$$\begin{pmatrix} 3 & -5 \\ 2 & -4 \end{pmatrix} \begin{pmatrix} x \\ y \end{pmatrix} = \begin{pmatrix} kx \\ ky \end{pmatrix}$$

Therefore

$$3x - 5y = kx \qquad \text{i.e. } (k - 3)x = -5y$$
$$2x - 4y = ky \qquad \text{i.e. } 2x = (k + 4)y$$
$$\frac{(k - 3)}{2} = \frac{-5}{(k + 4)}$$
$$(k - 3)(k + 4) = -10$$
$$k^2 + k - 2 = 0$$
$$(k + 2)(k - 1) = 0$$

Therefore

$$k = 1 \quad \text{or} \quad -2$$

These *two* values of k are called the *eigenvalues* of the (2×2) matrix $\begin{pmatrix} 3 & -5 \\ 2 & -4 \end{pmatrix}$.

$k = 1$ gives $2x - 4y = y$, i.e. $2x = 5y$, i.e. $\begin{pmatrix} x \\ y \end{pmatrix} = \begin{pmatrix} 5 \\ 2 \end{pmatrix}$

$k = -2$ gives $2x - 4y = -2y$, i.e. $2x = 2y$, i.e. $\begin{pmatrix} x \\ y \end{pmatrix} = \begin{pmatrix} 1 \\ 1 \end{pmatrix}$

In general the eigenvectors of the transformation are $\begin{pmatrix} 5m \\ 2m \end{pmatrix}$ and $\begin{pmatrix} m \\ m \end{pmatrix}$ where m is any scalar.

Now consider the general case.

What are the eigenvalues of the matrix $\begin{pmatrix} a & b \\ c & d \end{pmatrix}$?

$$\begin{pmatrix} a & b \\ c & d \end{pmatrix} \begin{pmatrix} x \\ y \end{pmatrix} = k \begin{pmatrix} x \\ y \end{pmatrix}$$

i.e.

$$\mathbf{A} \begin{pmatrix} x \\ y \end{pmatrix} = k\mathbf{I} \begin{pmatrix} x \\ y \end{pmatrix}$$

or

$$(\mathbf{A} - k\mathbf{I}) \begin{pmatrix} x \\ y \end{pmatrix} = 0$$

$$\left\{ \begin{pmatrix} a & b \\ c & d \end{pmatrix} - k \begin{pmatrix} 1 & 0 \\ 0 & 1 \end{pmatrix} \right\} \begin{pmatrix} x \\ y \end{pmatrix} = 0$$

$$\begin{pmatrix} a - k & b \\ c & d - k \end{pmatrix} \begin{pmatrix} x \\ y \end{pmatrix} = 0$$

Therefore

$$(a - k)x + by = 0$$

$$cx + (d - k)y = 0$$

The consistency of these equations requires

$$\frac{a - k}{c} = \frac{b}{d - k}$$

i.e.

$$\Delta = \begin{vmatrix} a - k & b \\ c & d - k \end{vmatrix} = 0 \qquad (9.6)$$

i.e. x and y have non-trivial solutions only when $|A - kI| = 0$.

Thus the eigenvalues of A are the roots of the equation obtained by putting the determinant of the matrix $(A - kI)$ equal to zero, i.e.

$$k^2 - (a + d)k + ad - bc = 0$$

This is called the *characteristic equation* of A.

9.13 The Eigenvalues of any Matrix

Let $X = \begin{pmatrix} x_1 \\ x_2 \\ x_3 \end{pmatrix}$ be a vector in three dimensions. Then

$$\lambda X = \begin{pmatrix} \lambda x_1 \\ \lambda x_2 \\ \lambda x_3 \end{pmatrix}$$

where λ is a real scalar, is a vector having the same direction as X.

Analogously, if X is a vector in n-dimensions we say that λX and X have the same direction.

Consider the matrix equation

$$AX = \lambda X$$

where A is a square matrix of order n and X is a column matrix of order $(n \times 1)$.

$$(A - \lambda I)X = 0$$

where I is a unit matrix of order n.

This equation has non-trivial solutions if and only if

$$|A - \lambda I| = 0$$

Therefore λ must satisfy the characteristic equation

$$
\begin{vmatrix}
a_{11} - \lambda & a_{12} & a_{13} & \cdots & a_{1n} \\
a_{21} & a_{22} - \lambda & a_{23} & \cdots & a_{2n} \\
a_{31} & a_{32} & a_{33} - \lambda & \cdots & a_{3n} \\
\cdot & \cdot & & & \\
\cdot & \cdot & & & \\
\cdot & \cdot & & & \\
a_{n1} & a_{n2} & a_{n3} & \cdots & a_{nn} - \lambda
\end{vmatrix} = 0 \tag{9.7}
$$

which leads to n values of λ which may be real or complex and need not all be distinct, i.e. to n eigenvalues which, in turn, give n eigenvectors which may be real or complex and may not be distinct.

EXAMPLE 9.2

Find the eigenvalues and the corresponding eigenvectors of the matrix

$$
A = \begin{pmatrix}
2 & 1 & 1 \\
2 & 3 & 4 \\
-1 & -1 & -2
\end{pmatrix}
$$

The eigenvalues λ are given by the matrix equation

$$AX = \lambda X$$

and are therefore the roots of the characteristic equation

$$|A - \lambda I| = 0$$

i.e. of

$$
\begin{vmatrix}
2 - \lambda & 1 & 1 \\
2 & 3 - \lambda & 4 \\
-1 & -1 & -(2 + \lambda)
\end{vmatrix} = 0
$$

Subtracting the second column from the first and the third column from the second, we have

$$\begin{vmatrix} 1-\lambda & 0 & 1 \\ -(1-\lambda) & -(1+\lambda) & 4 \\ 0 & 1+\lambda & -(2+\lambda) \end{vmatrix} = 0$$

$$(1-\lambda)(1+\lambda)\begin{vmatrix} 1 & 0 & 1 \\ -1 & -1 & 4 \\ 0 & 1 & -(2+\lambda) \end{vmatrix} = 0$$

$$(1-\lambda)(1+\lambda)(\lambda-3) = 0$$

$$\lambda = 1, -1 \quad \text{or} \quad 3$$

For $\lambda = 1$, the corresponding eigenvector is given by

$$\mathbf{AX} = \mathbf{X}$$

i.e.

$$\begin{pmatrix} 2 & 1 & 1 \\ 2 & 3 & 4 \\ -1 & -1 & -2 \end{pmatrix}\begin{pmatrix} x_1 \\ x_2 \\ x_3 \end{pmatrix} = \begin{pmatrix} x_1 \\ x_2 \\ x_3 \end{pmatrix}$$

Therefore

$$x_1 + x_2 + x_3 = 0$$
$$x_1 + x_2 + 2x_3 = 0$$
$$x_1 + x_2 + 3x_3 = 0$$

so that $x_1 + x_2 = 0$ and $x_3 = 0$.

$$x_1 : x_2 : x_3 = 1 : -1 : 0$$

This vector is arbitrary to the extent of an undetermined multiplier. Similarly, when the eigenvalues are $\lambda = -1$, 3 the corresponding eigenvectors are respectively given by

$$x_1 : x_2 : x_3 = 0 : 1 : -1$$
$$x_1 : x_2 : x_3 = 2 : 3 : -1$$

EXERCISES 9.3

1. Find the eigenvalues and eigenvectors of the following matrices:

(a) $\begin{pmatrix} 4 & -1 \\ 2 & 1 \end{pmatrix}$ (b) $\begin{pmatrix} 3 & \frac{5}{2} \\ -1 & -\frac{1}{2} \end{pmatrix}$ (c) $\begin{pmatrix} 3 & -4 \\ 1 & -1 \end{pmatrix}$

$$\left[(a)\ 2, 3;\ \begin{pmatrix} m \\ 2m \end{pmatrix}, \begin{pmatrix} m \\ m \end{pmatrix}\quad (b)\ 2, \tfrac{1}{2};\ \begin{pmatrix} -5m \\ 2m \end{pmatrix}, \begin{pmatrix} -m \\ m \end{pmatrix}\quad (c)\ 1;\ \begin{pmatrix} 2m \\ m \end{pmatrix} \text{ only} \right]$$

2. Comment on the possibility of eigenvalues for

(a) $\frac{1}{2}\begin{pmatrix} 1 & -\sqrt{3} \\ \sqrt{3} & 1 \end{pmatrix}$

(b) $\begin{pmatrix} 1 & -1 \\ 1 & 1 \end{pmatrix}$

3. Verify that the matrix

$$\begin{pmatrix} 2 & 1 & 0 \\ 1 & 2 & 1 \\ 0 & 1 & 2 \end{pmatrix}$$

has eigenvalues $2 \pm \sqrt{2}$ and 2 and corresponding eigenvectors proportional to $(1, \pm\sqrt{2}, 1)$ and $(1, 0, 1)$ respectively.

4. Show that the eigenvalues of the matrix

$$\begin{pmatrix} 2 & 0 & 1 \\ 0 & 3 & 0 \\ 1 & 0 & 2 \end{pmatrix}$$

are 3 (twice) and 1. Show that the eigenvector corresponding to 1 is $k(1, 0, -1)$ and that corresponding to 3 is any vector of the form $k'(1, 1, 1)$, where k, k' are arbitrary constants.

Index